How Oil and Gas Companies are Pivoting Toward Climate Change

How Oil and Gas Companies are Pivoting Toward Climate Change

By

Ian Palmer

Cambridge Scholars Publishing

How Oil and Gas Companies are Pivoting Toward Climate Change

By Ian Palmer

This book first published 2024

Cambridge Scholars Publishing

Lady Stephenson Library, Newcastle upon Tyne, NE6 2PA, UK

British Library Cataloguing in Publication Data
A catalogue record for this book is available from the British Library

ISBN (10): 1-0364-0563-X
ISBN (13): 978-1-0364-0563-2

For Mary Ann and Kim—for your enthusiastic support during the tough life challenges we all endured while this book was being written. Your resilience has been an inspiration to me.

CONTENTS

FOREWORD

This book, *How Oil and Gas Companies Are Pivoting toward Climate Change*, was an enjoyable read. It is written using clear and understandable prose with topics well researched and explained.

The topics presented show how the largest Western-world oil and gas companies are dealing with the realities of fossil fuel–induced global warming and their resulting climate changes while still maintaining profitability.

The book draws the distinction between US and European major oil companies, where US companies have strategies of maintaining and sometimes actually increasing production with plans to compensate for CO_2 emissions using CCS [carbon capture and storage] or DAC [direct air capture], and to reform methane to hydrogen fuel. In contrast, European majors are actually planning to replace some of their oil and gas production with wind, solar, and geothermal.

The book gives a very nice overview of how LNG [liquefied natural gas] is like liquid gold and has been ramped up to aid Europe and Asia. The description of how Occidental Petroleum, Chevron, and ExxonMobil + Pioneer are operating in the Permian basin shows that US oil and gas production will be alive and well for at least the next twenty-five years.

As to CO_2 abatement and storage, the book shows how oil and gas companies are cooperating with governments in working to achieve net-zero CO_2 emissions by 2050. However, this will not be easy.

With regard to renewable energy, the book gives a very pertinent example of Australia as a success story in transition to renewable energies. It explains how solar, wind, and batteries are replacing coal, oil, and gas but stresses how the world's transition has a long way to go, especially in the US.

As for possible legal problems during this transition to renewables, the book gives the case of a California lawsuit versus big oil. The story shows how difficult it will be to hold big oil (e.g., ExxonMobil) responsible for climate change, even though they predicted very accurately the rise of global temperatures from 1980 up to 2020 when fossil fuel production was included.

The book states that global warming is not the same as climate change and that such damaging elements of climate change (the "killer quad" of droughts, floods, hurricanes, and wildfires) have not been shown to change significantly in the last forty to fifty years. The book says this is consistent with a lack of "climate signals" in the IPCC AR6 [Sixth Assessment Report] (ca. 2023).

The book also says, "If extreme weather events haven't shown any worsening over the past forty to fifty years, when global temperature rose by almost 1°C, there should be less urgency about temperatures rising another 0.3°C to 1.5°C, which the Paris Agreement wanted to prevent by 2050."

I like the way the book ended with a chapter citing other forms of energy that could be developed such as liquid biofuels, fission, fusion, and geothermal. Also, the table from Bill Gates showing additional technologies adapted to reduce GHG [greenhouse gas] is instructive.

In conclusion, the book gives a wealth of topical insight and information that readers can gain to help them understand the climate and geopolitical issues of today's world. The book will help anyone who wants to be informed about climate change and the energy transition but does not have the time or inclination to wade through massive documents on the subject produced by the IPCC, the IEA, BP, etc.

John Cameron
John R. Cameron, with a PhD in ChE from the University of Michigan, has worked over 38 years in the upstream oil and gas industry as an engineer doing research and consulting for major and independent oil companies, service companies and universities.

PREFACE

The hard facts are as follows. The US Energy Information Administration (EIA) has forecast US oil production to reach almost thirteen million barrels per day in 2024 and natural gas to rise to 105 billion cubic feet per day—both new records.

Global greenhouse gas emissions hit a (projected) record high in 2023 of 37.6 Gt of CO_2 equivalent from fossil fuels and industry (Statista 2023a).

In 2023, the world spent a record \$1.7 trillion on clean energy, according to Bloomberg, which surpassed the \$1.1 trillion spent on fossil fuels. Surprisingly, solar and electric cars in China dominated the difference.

The oil and gas industry employs twelve million workers across the world and supplies over half of the global energy consumed. But the industry's investment in clean energy amounts to just 1% of the worldwide investment. Capital spending on clean energy by oil and gas has been only 2.5% of the industry's total capital expenditures. It would need to be 50% by the year 2030 to reach the goals of the Paris Agreement of 2015.

What should the oil and gas industry do? Start to phase out oil and gas? Or scale up technologies like carbon capture and storage or hydrogen to compensate for business-as-usual oil and gas drilling?

With the unique perspective of a solar space scientist turned oilfield engineer, the book takes an honest look at the fast-changing energy transition.

- How far are global leaders pivoting toward climate change? BP? Occidental? ExxonMobil?
- ExxonMobil confirmed global warming forty years ago. Did they really try to deceive people about climate change?
- Are CCS and hydrogen fuel all they are cracked up to be?
- Batteries are surging, and are displacing gasoline (oil) in cars and trucks and also coal and natural gas in power plants.
- Extreme weather events have not worsened with time over the past forty years, when global temperature has gone up by almost 1°C. So, is the oil and gas industry not responsible for these killer extreme events?

- Is climate change urgency as dire as alarmists are saying? If it's not, will LNG export restrictions be lifted?
- Can oil and gas companies reach net-zero carbon emissions by 2050 without new technology?

While the oil and gas industry confronts challenging choices in regard to climate change, this book also reaches down to a personal level. The book is a handy reference to what is going on both inside and outside the oil and gas business that will undoubtedly impact people working in the business. It is good to keep an eye on the weather!

ACKNOWLEDGMENTS

Sincere thanks to Margaret Rogalski, who parlayed my earlier book *The Shale Controversy* to the rest of the world and opened the door to Forbes.com.

Special thanks to Chris Helman, editor, who took a chance and gave me the opportunity to write for Forbes.com. I am indebted to Forbes.com for the many opportunities to contribute.

Deep thanks to Gillian Felix, who has for several years designed and promoted my writings, professional and personal, in books, on my website, www.IanDexterPalmer.com, and on LinkedIn.

Mindi Machart edited my manuscript with professional dexterity and interpersonal cordiality.

John Cameron and Bob Moore perused an early version of the manuscript and provided comments, and John kindly provided the foreword to this book.

INTRODUCTORY NOTE

The oil and gas industry is wrestling with an energy transition that is linked to climate change. These concepts are very complicated, especially when government is linking up with industry. This book describes how oil and gas companies are adapting their thinking and making changes in their operations as they pivot toward climate change. While not a comprehensive study, many different facets of the transition are addressed in this book, and these lead to learnings that will help define the way forward for the oil and gas industry.

LIST OF ABBREVIATIONS

AR6	Sixth Assessment Report
AREH	Australian Renewable Energy Hub
Bcfd	billion cubic feet per day
bcm	billion cubic meters
boepd	barrels of oil equivalent per day
CapEx	capital expenditures
CCS	carbon capture and storage
CH_4	methane
CO_2	carbon dioxide
COP	Conference of the Parties
CSIRO	Commonwealth Scientific and Industrial Research Organisation
DAC	direct air capture
DOE	Department of Energy
EDF	Environmental Defense Fund
EIA	Energy Information Administration
EJ	exajoules
EOR	enhanced oil recovery
EPA	Environmental Protection Agency
G-7	Group of Seven
GDP	gross domestic product
GHG	greenhouse gas
Gt/yr	gigatons per year
GW	gigwatt
GWh	gigawatt hours
HERO	Hydrogen Energy Release Optimiser
IEA	International Energy Agency
IPCC	Intergovernmental Panel on Climate Change
IRA	Inflation Reduction Act
LNG	liquefied natural gas
MMboepd	millions of barrels of oil equivalent per day
MMbpd	million barrels per day

MMbtu	million British thermal units
MW	megawatt
MWh	megawatt hours
NGLs	natural gas liquids
NOAA	National Oceanic and Atmospheric Administration
NSW	New South Wales
OPEC	Organization of the Petroleum Exporting Countries
Oxy	Occidental Petroleum Corporation
ppm	parts per million
PV	photovoltaic
RCSPs	regional carbon sequestration partnerships
RNG	renewable natural gas
SMR	small modular reactor
Tcf	trillion cubic feet
tpy	metric tons per year
TW	terrawatts
UAE	United Arab Emirates
UIC	Underground Injection Control
UN	United Nations

CHAPTER 1

PUT YOUR MONEY ON THE PERMIAN: THE PHAR LAP OF OIL AND GAS BASINS

The Permian basin is one of the world's premier oil and gas basins. It has been projected that Permian crude oil will grow steadily from about 6 MMbpd in mid-2023 to possibly 8 MMbpd by 2030. Meanwhile, total US production may reach 13.3 MMbpd by 2050.

Phar Lap was a champion racehorse bred in New Zealand and raced in Australia. He won thirty-seven races from fifty-one starts, including the 1930 Melbourne Cup, and was an inspiration to people in the early years of the Great Depression. He died from lead poisoning in the US, but the cause was controversial. His heart was 50% heavier than the average for racehorses, and it was given to the National Museum in Australia. Phar Lap and Secretariat have been called the two greatest racehorses in the world.

The Permian basin is a bit like Phar Lap. It is winning everywhere and is the center of attention for the US and the rest of the world. In 2023, ExxonMobil, for example, bought Pioneer Natural Resources for a huge price, and the merged company will have the largest oil and gas production in the Permian basin—about 11% of the total basin production. All this adds up to a gargantuan company with enormous influence over oil and gas but also over climate change.

The Permian Races Ahead of Pandemic Lows

In early 2022, the Permian basin in West Texas and Southeast New Mexico produced 5 million barrels per day (MMbpd) of crude oil, which was almost half of the total US supply of crude oil (Palmer 2022a; Mathews 2023a). It was projected at the time that the US had seventy-six billion barrels of untapped reserves, and most of this lay under the Permian (Gaswirth et al. 2018).

The peak of crude oil production in the Permian before the 2020 pandemic was 4.9 MMbpd. But production in the Permian rose to a record 6.0 MMbpd by end of 2023. The Permian is still running hard.

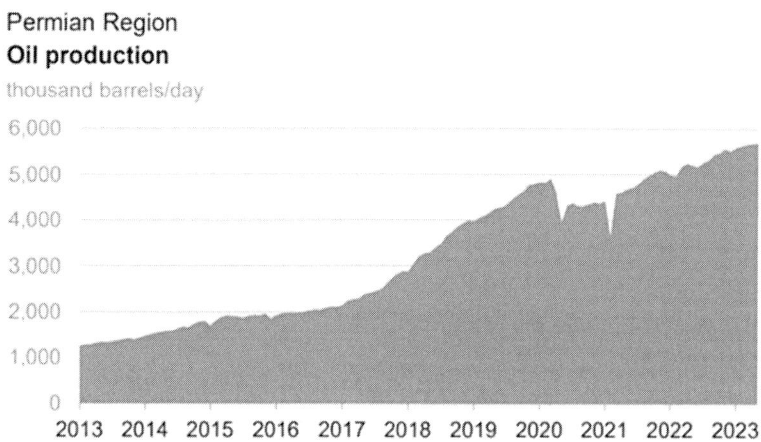

Fig. 1-1. Crude oil production from Permian basin (Mathews 2023a).

Other shale-type producers are increasing too. The Bakken in North Dakota and Montana reached 1.2 MMbpd in 2023 but remained a good bit lower than its peak of 1.54 MMbpd, recorded in November 2019. In early 2023, the Eagle Ford in Texas rose to nearly 1.2 MMbpd but also lay well below its pre-pandemic peak of 1.40 MMbpd (EIA 2024a).

Natural gas production in the US has been surging even more. Associated gas in the Permian rose from 17.4 billion cubic feet per day (Bcfd) pre-pandemic to about 22.5 Bcfd in May 2023—a surge of 29%. Total production in the US was about 100 Bcfd, to which the Permian contributed a remarkable 23%.

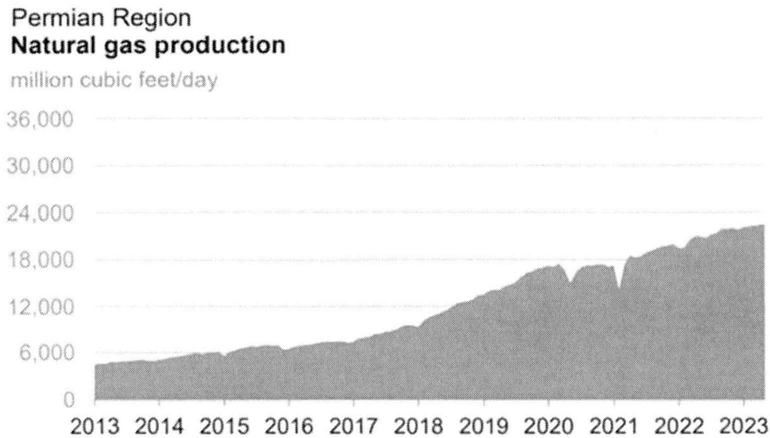

Fig. 1-2. Natural gas production from Permian basin (Mathews 2023a).

In the Haynesville Shale, which straddles the border between Louisiana and Texas, natural gas production rose from 12.0 Bcfd pre-pandemic to 16.8 Bcfd in May 2023. This was a surge of 40%. The largest gas play in the US, the Appalachian—which includes the Marcellus Shale—actually fell from 36 Bcfd pre-pandemic to 35.3 Bcfd in May 2023.

What are the Keys of Success in the Permian?

In late 2018 the United States Geological Survey completed an assessment of the entire Delaware basin, a subset of the Permian existing in both Southeast New Mexico and Far West Texas (Gaswirth et al. 2018). They came up with over forty-six billion barrels of technically recoverable continuous oil resources, plus 281 trillion cubic feet (Tcf) of natural gas and twenty billion barrels of natural gas liquids (NGLs). NGLs are liquid compounds more complex than methane (the major component of natural gas), such as ethane and pentane.

What to remember: The Permian is a catchword for wealth and energy security for the US. But what will happen to it during the energy transition?

The immense quantities of oil and gas in the Delaware basin make up the largest deposit of oil and gas ever documented by the United States Geological Survey in the US. Quite simply, it's the nation's dominant energy play, with some of the largest recoverable reserves in the world.

The technical key is a few horizontal wells drilled per pad in stacked pay zones, each horizontal about two miles long and fracked up to forty separate times along its length. Monster wells making more than 10,000 barrels of oil equivalent per day (boepd) have been reported (Palmer 2020a).

Permian Will Peak at 8 MMbpd in 2030

The betting on the Permian future is positive. Five watchdog heavyweights forecast crude oil production rising steadily through 2028 before peaking in the range of 7.3–8.0 MMbpd around 2030 (Crowley, Wethe, and Ferman 2023). The watchdog companies are the Energy Information Administration (EIA), Wood Mackenzie, Rystad, Enverus, and S&P Global.

According to a Bloomberg survey, oil production will expand by 40% through 2030 (Crowley, Wethe, and Ferman 2023). After plateauing for a few years, a slow decline will set in in 2035, which promises oil will be a consistent supplier of energy consumption for decades.

The Permian will offset production declines from other shale basins in the US. Federal forecasts suggest total US oil production stabilizing in the range of 12.3–13.3 MMbpd through 2050.

US oil production in 2022 was 11.9 MMbpd, about 12% of global production (99.8 MMbpd), and looks like a stayer and a good bet to last the distance until 2050. The US produces more oil than Saudi Arabia (No. 2) and Russia (No. 3). In the first half of 2023, US oil exports reached a record of 4 MMbpd—not bad for a country that was prohibited from exporting oil until 2015.

The Permian Feeds LNG Futures

As of mid-2023, the Permian and Haynesville have increased their natural gas production by 4–5 Bcfd since the pandemic of 2020. This is a surge of 29% for the Permian and 40% for Haynesville.

It's not a coincidence that both these resources lie close to the liquefied natural gas (LNG) export terminals spread along the Gulf of Mexico. LNG has a very bright future (Palmer 2023a).

Since the war in Ukraine, investments in LNG have moved quickly from uncertainty due to supply-chain delays and climate-fearing stranded assets to a position of "Let's get the funding and permitting done as soon as possible" (Palmer 2023b). US contractors in 2023 have been arm-wrestling over long-term exports with Europe, China, and Southeast Asia countries like Japan and South Korea.

Takeaways

The peak of crude oil production in the Permian before the pandemic was 4.9 MMbpd. But production in the Permian rose to 6.0 MMbpd by the end of 2023. Five watchdog heavyweights forecast crude oil production in the Permian rising steadily through 2028 before peaking in the range of 7.3–8.0 MMbpd around 2030.

US oil production in 2022 was 11.9 MMbpd, about 12% of global production, and looks like a good bet to last the distance until 2050. Federal forecasts suggest total US oil production will stabilize in the range of 12.3–13.3 MMbpd through 2050.

Associated gas in the Permian has risen from pre-pandemic 17.4 Bcfd and reached 22.5 Bcfd in May 2023—a surge of 29%. The golden age for LNG has a strong connection with the Permian. US contractors are currently arm-wrestling over long-term LNG exports with Europe, China, and Southeast Asia countries like Japan and South Korea.

The Permian basin is the capstone of US oil and gas basins, with its production about half and gas production about a quarter of total US production.

CHAPTER 2

VICKI HOLLUB AND JOE BIDEN: UNLIKELY TIES BETWEEN ENERGY SECURITY AND ENERGY TRANSITION

This chapter looks at two important influencers in the field of energy and finds they have common interests that are surprising.

Vicki Hollub is CEO of Occidental Petroleum Corporation. She was the first woman CEO of a large US oil and gas company. Joe Biden is president of the United States.

Previously, a short answer to the question "What do Vicki Hollub and Joe Biden have in common?" would have been "Not much," even though Hollub did have a past connection with Washington politics: she was an advisor to the US Department of Energy during the Trump administration.

Joe Biden came into office in 2020 with plans to stall further expansion of oil and gas, on the belief that renewable energies were the future and oil and gas would fade away. Biden canceled the Keystone XL expansion on his first day in office. Biden paused the leasing of future oil and gas wells on federal land soon after that.

Conversely, Occidental (Oxy) purchased Anadarko in 2019 for a pricey $55 billion, beating out Chevron in the process. Hollub's commitment to the future of oil and gas was obvious.

Hollub would have been aware that the oil and gas industry adds $1.8 trillion to the US economy each year (7.6% of gross domestic product) and supports nearly eleven million jobs (5.4% of total employment) (Jacobs 2023a).

Unexpected Changes

But in the past two years things have changed in unexpected ways.

First came COVID-19 in 2020, and the price of oil collapsed, actually dropping to zero for a short time. It must have been a scary time for Oxy so soon after their Anadarko purchase. But the rebound after COVID-19 was strong for the general economy and for the oil and gas industry in particular.

Then came the Russian invasion of Ukraine in February of 2022. Russia cut gas supplies to Europe, and gas prices hit the roof. Oil prices zoomed up in concert as economic sanctions hit Russia, some countries stopped buying their crude oil, and an oil price cap was instigated by the EU that reduced Russia's sales of crude oil.

Oil sales by Russia were at 23% of the total Russian budget in mid-2023, down from 30%–35% of the Russian budget before the start of the war in Ukraine (Meredith 2023).

But crude prices remained high, averaging $100 per barrel in 2022. As a result, oil and gas companies made record profits in 2022—and the profits were twice as much as the previous record year of 2018 (King 2023).

In the years since 2000, Europe had stepped away from oil and gas and coal while supporting renewables. But renewables could not fill the gap caused by the Russian cuts.

Energy Security

From this European experience, what became clear to Vicki Hollub and Joe Biden was that, first, energy security was critically important to a country or a group of countries with close economic ties, such as the EU.

Second, the US could assist Europe by providing oil and gas, especially LNG, to replace lost Russian supplies. In the twenty-first century, Europe walked away from their own oil and gas industry, preferring instead to rely on oil and gas from Russia. In contrast, Hollub said:

> "In terms of energy independence, hydraulic fracturing has helped the United States maintain a position of power in the world," she said, adding that the major shale plays developed with fracturing account for around 8 [MMbpd] out of the total of 11.6 [MMbpd] of crude that the US produces today. (Jacobs 2022)

The US became self-sufficient in oil and gas production in 2020 for the first time in seventy years, and this can be attributed to the shale and fracturing revolution that began around 2003 (Palmer 2023a). This has enabled the US to open up its exports of LNG to help Europe meet its energy needs after the invasion of Ukraine. The US is now close to being the largest exporter of LNG in the world.

Hollub spoke further of the geopolitical advantages the oil and gas industry has given the US:

> "I know how [our policymakers] use or leverage" US energy independence in tandem with its defense capabilities, she said. "When you put the two of those together, there's no country in the world that can really challenge us very much" (Jacobs 2022).

These statements by Vicki Hollub could have been made by President Biden, because they refer to integral parts of US national security and its assistance to the EU while the Russian invasion continues.

The Energy Transition

Self-sufficient energy and energy security are not the only things that Vicki Hollub and Joe Biden have in common. They both have a commitment to energy transition by lowering or removing or offsetting greenhouse gas (GHG) emissions—on the way to the Paris 2015 goal of net-zero emissions by 2050.

Oxy's emissions come in three types: Scope 1, 2, and 3. Scope 1 and 2 fall under the responsibility of Oxy, but Scope 3 are emissions by consumers—companies or people who burn the oil and gas or refined products sold by Oxy.

Oxy has pledged to offset Scope 3 emissions, which is radical for an oil and gas company because Scope 3 can be 95% of total emissions related to the company (Jacobs 2022; Palmer 2022b). It's not only the largest source of emissions; it's also the most difficult to control or reduce.

Whoever adopts the blame for Scope 3 emissions—producer or consumer—it's a dominant part of the 73% of global emissions that comes from burning fossil fuels such as oil or gas or coal. Understanding Scope 3, then, is essential to move beyond measuring emissions during production to understanding the total impact of doing business.

For Oxy, the willingness to include Scope 3 emissions makes the company an outlier. The strategy is to use their expertise in carbon capture and storage (CCS), which is a powerful way to offset emissions, for themselves or even for other companies. Basically, Oxy hopes to bury enough CO_2 to compensate the CO_2 that will be released by burning their oil-gasoline in cars and by burning their natural gas in electrical power plants.

Oxy's Carbon Capture Plans

Like President Biden, Vicki Hollub believes in climate change and Oxy's need to do something ambitious about it: "As we started to learn more about climate change and climate mitigation, we realized this plan needed to happen, and would happen, to help the world" (Jacobs 2022).

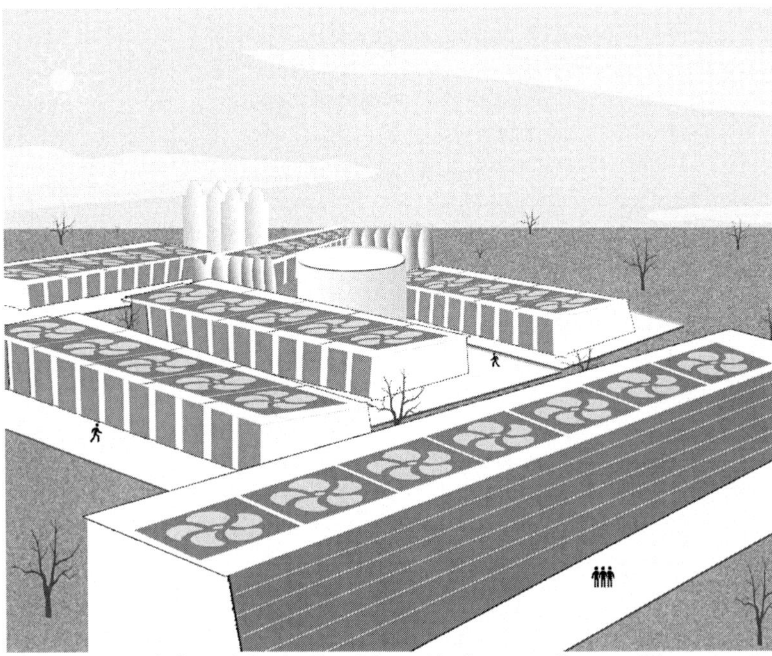

Fig. 2-1. A rendering of Oxy's first Direct Air Capture—Stratos. (Illustration by Gillian Felix.)

Oxy has big plans for direct air capture (DAC), a new technology that uses a wall of fans to blow air over chemicals that suck CO_2, the main GHG, out of the airstream (Palmer 2022c).

The first scaled project, called Stratos, should come online in 2025 in West Texas. A $1.3 billion project, this will separate out CO_2 at the rate of 500,000 metric tons per year (tpy). Other facilities are planned to separate

1,000,000 tpy. This is only 0.016% of total US emissions, but Oxy's goal of one hundred such plants would remove 1.6% of US emissions.

The world's largest investment manager, BlackRock, will invest $550 million in Stratos (Valle 2023). ADNOC (Abu Dhabi National Oil Company) will fund a DAC study with Oxy in United Arab Emirates for a 1,000,000 tpy facility.

Oxy envisions a franchise type of business model, where they charge a licensing fee and the plant is managed by partners. Oxy foresees one thousand such plants across the world will be needed by 2050. If the potential is one thousand such plants, this would equate to removing about 2% of current global emissions. The numbers show how difficult it will be to reach net-zero emissions by 2050.

DAC will be a significant part of carbon capture, but not a dominating part, according to the International Energy Agency (IEA). Carbon will be captured from many sectors of the energy economy (in order), including natural gas, coal, DAC, biomass, and hard-to abate areas. By 2050, IEA predict natural gas at the top, DAC in the middle, and hard-to abate areas, at the bottom, totaling 6.6 Gt per year (IEA n.d.).

Carbon removal credits will play a big part in all this. It's hard to believe, but Oxy has already sold CO_2 removal credits to the Houston Texans football team to offset their emissions from air travel for three seasons. Other companies who have pre-purchased such credits include Amazon, Airbus, and Shopify. The market for carbon removal credits may eventually surge to hundreds of billions of dollars (L. 2023).

Hollub pointed out that a competitor called Climeworks, in Iceland, is capturing only 4,000 tpy, and the method is difficult to scale up to much greater volumes.

CERAWeek

Another thing in common between Hollub and Biden: tax breaks are a big incentive in the congressional Inflation Reduction Act (IRA) for investments in DAC and CCS projects.

In March 2023, John Kerry spoke at CERAWeek to a room packed with oil and gas CEOs. Kerry is the president's climate czar. Kerry and the president sent a message: "We need to not be polarized here. We need to collaborate" (Worland 2023).

The subject was the IRA bill. The bill funds wind and solar renewables but also (1) provides tax incentives for CCS, (2) subsidizes hydrogen, and (3) supports biofuel production. The promise of billions of dollars changes the math for the oil and gas industry, because the industry already has know-how and skills to build these enterprises to scale, and now they have ten years of IRA carrots to take to the bank.

What to remember: Energy security and the energy transition need to grow together.

The industry is seeing the opportunities and flocked to sessions at CERAWeek to hear about the Fed's initiatives. After decades of supporting environmental progress and recent opposition to improving oil and gas enterprises, the federal government is wanting to include oil and gas in its climate change program. "If they want to play clean, we welcome them to the party," said John Podesta, the Biden advisor who is leading IRA implementation (Worland 2023). It seems like Vicki Hollub would shake Joe Biden's hand and say with a smile, "Count us in."

And a bunch of oil and gas people would have liked to shake Biden's hand when the president announced that Conoco-Phillips could drill wells in the Willow project in the National Petroleum Reserve-Alaska (Newman 2023).

Biden is learning that the transition to green energies is a *both/and* proposition. The US needs continued access to oil and gas while developing renewable energies that will gradually replace them. But at the same time oil and gas needs to offset leftover carbon emissions to satisfy net-zero emissions by 2050.

Federal forecasts suggest total US oil production will stabilize in the range of 12.3–13.3 MMbpd through 2050. But it's not clear whether switching from gasoline to electric vehicles, which will have a dramatic effect on oil production, has been included in this prediction (Palmer 2021a).

For the past two years, Vicki Hollub has been saying that Oxy is becoming a carbon-management business. What they perceive as a major business opportunity is starting to take shape.

Takeaways

Though surprising, Vicki Hollub and President Biden have a lot in common right now: the price of oil, energy security, the energy transition, carbon capture and storage (CCS), and direct air capture (DAC). And, last but not least, attractive contract monies and tax credits newly minted by the US government to develop endeavors that reduce GHG emissions.

CHAPTER 3

THE PROFIT DILEMMA FOR FOSSIL FUEL COMPANIES: ENERGY SECURITY VERSUS CLIMATE SECURITY

This chapter discusses a way for countries to assess their energy and climate security, to make a comparison with other countries, and even to create new policy.

In recent years, energy security has been a source of anxiety across the world. Inflation was high, including the cost of natural gas and electricity, and many had predicted a worldwide recession.

Things in Europe were bad because of the cuts in oil and gas triggered by the Russian invasion of Ukraine. And wars in Ukraine and Gaza were still going on at the beginning of 2024.

The United Nations (UN) gather once a year at a Conference of the Parties (COP) meeting, where international reps discuss climate security. A common refrain is that not enough progress is being made to reduce the GHG emissions that cause global warming and ensuing climate disasters.

On one hand the world faces energy insecurity, while on the other it faces climate insecurity. It's a great irony that the fossil fuel industry is caught in the middle, because it is responsible for about 83% of the world's energy and about 73% of global carbon emissions. The oil and gas industry provides 57% of the world's energy—and about 50% of the world's emissions (Hausfather 2021.)

For oil and gas companies, 25% of their emissions comes from their production and distribution of oil and gas products. But 75% of their emissions comes from third parties burning their oil and gas products (in cars, gas-fired power plants, commercial and residential gas heaters, etc.).

It's insightful to compare energy security versus climate security, country by country, on the same page. This information can be gleaned from data discussed by the World Energy Council (2022). This data was presented in a table of countries, which has been simplified in table 3-1 (Palmer 2022d).

If a country scores low in *energy security* proxy and *climate security* proxy, then that country is rated highly for its approach to energy security *and* climate security. Note that the *energy security* proxy list does not begin with a rank of one because it's an average of two separate rankings.

If a country's energy security proxy is a high number and its climate security proxy is a low number, this country scores well on climate security but low on energy security.

Table 3-1 only includes results for a short list of countries out of the full list of 120 countries.

The bottom line is Sweden is ranked at the top, the UK is second, Canada is sixth, the US is thirteenth, Australia is fourteenth, and China is twenty-second.

The last two columns in table 3-1 show (1) a sum of the two proxies and (2) the country ranking. In general, as we move down the table, both energy security and climate security decrease and become a reason for more concern.

Country	Energy security proxy	Climate security proxy	Index sum	Ranking based on index sum	
Sweden	23	1	24	**1**	**Sweden**
Denmark	25	4	29	**2**	**Denmark**
UK	22	7	29	**2**	**UK**
Austria	22	11	33	**4**	**Austria**
Switzerland	34	2	36	**5**	**Switzerland**
Canada	13	26	39	**6**	**Canada**
Finland	26	13	39	**6**	**Finland**
France	35	6	41	**8**	**France**
Germany	26	16	42	**9**	**Germany**
New Zealand	33	18	51	**10**	**New Zealand**
Norway	52	3	55	**11**	**Norway**
Spain	46	15	61	**12**	**Spain**
US	11	56	67	**13**	**US**
Australia	35	47	82	**14**	**Australia**
Italy	52	31	83	**15**	**Italy**
Netherlands	50	37	87	**16**	**Netherlands**
Brazil	79	8	87	**16**	**Brazil**
Japan	68	32	100	**18**	**Japan**
Russia	56	64	120	**19**	**Russia**
Saudi Arabia	59	97	156	**20**	**Saudi Arabia**
Mexico	118	42	160	**21**	**Mexico**
China	80	82	162	**22**	**China**
Qatar	54	108	162	**22**	**Qatar**
Indonesia	107	66	173	**24**	**Indonesia**
India	123	99	222	**25**	**India**
Nigeria	182	103	285	**26**	**Nigeria**

Table 3-1. Energy and climate security data in 2022. (Table modified by permission from Oliver Wyman and the World Energy Council; World Energy Council 2022.)

Country Assessments

Summaries are available for many but not all countries in table 3-1 (World Energy Council 2022). The World Energy Council's report shows that two Baltic countries rank highest, along with the UK. Their energy security is good but not as good as the US's or Canada's. Their climate security is very good, equaled only by Norway (another Baltic country) and France. Note that Norway exports large amounts of carbon emissions in the unburned form of oil and gas, but it's not clear if that was considered in the original table.

The UK is advanced in climate security. Since the year 2000, CO_2 releases have fallen to less than half. This is because coal usage was drawn down and renewable energies boosted the generation of electricity.

Canada is ranked sixth and is advanced in energy security. Its climate security improved in 2023 due to government policy.

France and Germany fall in the top ten. Their climate security is in good hands (France is better), but they don't score as well in energy security as the US and Canada, for example. The US and Canada have the benefit of the shale revolution to thank for this.

The US is ranked thirteenth. Its energy security is excellent because the US has a secure supply of energy with a mix of natural gas, coal, nuclear, hydropower, and renewables. Climate security is middle of the road due to the release of so much emissions, but 2024 should show improvement due to radical climate policy and oil majors adopting CCS as well as hydrogen production.

Australia is ranked fourteenth, one behind the US. The country is okay in energy security, with almost no reliance on energy imports. The country underperforms in climate security, but a new government in 2022 promised stronger action on climate plus rapidly expanding solar, wind, and battery developments. Surprisingly, the state of South Australia had 72% of its electricity supplied by renewable energies in 2023.

Middle East countries like Saudi Arabia and Qatar don't score as well as expected in energy security—probably due to their lack of diversification in electricity and storage of energy. Their positions in climate security are at the concerning end of the distribution.

Japan is middle of the road for an advanced economy. Its climate security is improving over time. But it needs to import less energy to improve its energy security.

Developing countries such as China, India, and Indonesia do not contain summaries in the article for some reason. But from table 3-1, it's clear that their poor scores in both energy and climate security reflect the challenges imposed by their immense populations.

The Profit Dilemma for Fossil Fuel Companies

There are a number of factors that influence the profit dilemma between energy security and climate security. World events have affected both energy and climate security, because they are locked together. Ukraine is an example. The Russian invasion intensified the dilemma for global oil and gas—while oil and gas were needed to support energy security in Europe, oil and gas contribute about 50% of global carbon emissions. As certain countries kicked up oil and gas production (and particularly LNG) to support Europe, emissions increased also.

Major oil company executives smiled as their profits reached an all-time high in 2022 when the price of crude oil averaged $100 per barrel due to energy insecurity (fig. 3-1). Employees smiled. Stockholders smiled. A vast number of workers in the oil and gas industry smiled because of job security. And Europe in 2023 was still buying from the US three times as much LNG as was Southeast Asia, including China.

In fig. 3-1, the 40,000 number in the chart actually represents $40 billion in profit. ExxonMobil, second from top in the chart, is an outsize winner. Its influence in the world of oil and gas is unmistakable. Also note that the profit uptick later in the chart followed a period of falling profits, which culminated in the pandemic years. In 2019 oil companies tightened up by cutting costs and curtailing their previous leasing and drilling expansions. The point was to increase their profit lines and, of course, their stock prices.

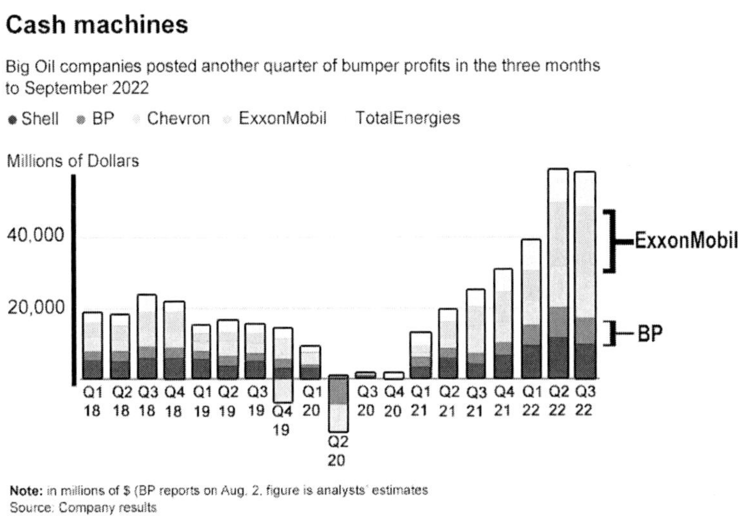

Fig. 3-1. Quarterly profits from oil and gas majors. (Photo courtesy of Reuters; Valle and Bousso 2022.)

In 2022 it became clear to oil companies, and the world, that energy security needs to be protected. The dilemma arose because only oil and gas energies, with their LNG, could fill the gap—but they cause about 50% of global carbon emissions. As a result, climate security got shifted to the back burner.

Hisham Zubari, senior advisor at the National Oil and Gas Authority in Bahrain, said one of the problems is that making a transition to renewables entails seven to ten years to pay back the investment. This is too long for profit-driven companies. "A balance needs to be sought between investment recovery and climate action. We [the world] don't have a model for this," he said (SPE 2021).

BP's experience: BP's shares gained more than 45% in 2022, buoyed by stronger oil and gas prices. But they trailed gains by Shell and US rivals ExxonMobil and Chevron—probably because BP had diversified more. BP's stated goal was 40% renewables by 2030, but in 2023 this was changed to 25% by 2030.

What to remember: Fossil fuels provide energy security that comes with company profits. If climate security means lower profits, this is a dilemma for the companies and maybe for the world.

When dissecting the profit motive, there is also an argument for the greater good of mankind. While profits are good for ExxonMobil, what about sea level rise in Vanuatu? Or retreating glaciers in the Third Pole region of the world? *Social license to produce oil and gas* is another name for it. Some ExxonMobil investors felt that with regard to the energy transition, the company was lagging behind other major oil companies.

As a stockholder, Chris James and his investment company, called Engine No. 1, proposed four new candidates (Baer and Lim 2021) for the board of ExxonMobil, all of whom had environmental experience. Three of them were elected. This was 2021 and seemed to be a factor in ExxonMobil making a U-turn in their attitude toward climate change.

One big agreement by countries at COP28 in December 2023 was as follows:

> For countries to quickly shift energy systems away from fossil fuels in a just and orderly fashion. . . Countries also are called to contribute to a global transition effort—rather than being outright compelled to make that shift on their own. (Ainger, Shankleman, and Dlouhy 2023)

This is in stark contrast to a UN study labeled the *Production Gap Report*, which states that twenty major producers of fossil fuels plan on production levels in 2030 that are double what is necessary to reach the global warming levels of 1.5°C set by the Paris Agreement of 2015 (Twidale 2023; SEI 2023). Instead, present global oil production of about 100 MMbpd would need to drop by 14%.

The truth is the world needs fossil fuels and will for a long time to come. If oil and gas production stopped suddenly, the world would starve, because food is transported by semitrailers or by rail or by ships, which mostly run on oil and gas.

At present, European oil and gas companies are continuing to diversify into renewables. But US companies are reluctant to diversify, preferring to invest in CCS and hydrogen to reduce emissions as well as fixing methane leaks and cleaning up their own operations' carbon footprint.

Tensions have always existed and still remain over the priorities of fossil energy security versus climate change. Many EU countries, and even oil majors, have given priority to renewable energies in the past twenty years. One lesson of 2022 is to be careful to preserve energy security when transitioning toward renewables.

Takeaways

Table 3-1 provides a way for countries to assess their energy security and climate security and even make a comparison with other countries. The analysis may be useful for governments to create new policy or stimulate old policy in respect of energy and climate security.

Table 3-1 points out the disjunct between countries like the US and Canada, which score well on energy security, and European countries like Sweden, Denmark, the UK, and France, which score well on climate security. Two Baltic countries and the UK have the best balance between energy and climate security.

In 2022 it became clear to the world that energy security needs to be protected because only oil and gas energies, such as LNG, could fill the gap created by the Russian war in Ukraine. But the dilemma is that oil and gas companies and their products cause about 50% of carbon emissions.

BP has led in diversifying to renewables. BP's stated goal was 40% renewables by 2030, but in 2023 this was changed to 25% by 2030. But ExxonMobil firmly changed direction in 2021, as this book will reveal, and the company has become a world influencer—not only in oil and gas, but also in climate change.

CHAPTER 4

THE GOLDEN AGE FOR LIQUEFIED NATURAL GAS (LNG)

Within two years, LNG replaced all the lost gas from Russia. US export revenues of LNG grew exponentially through 2022. Export volumes and wealth from LNG could potentially lead to $100 billion in new LNG developments in the US.

The importance of energy security on a geopolitical stage is clear. The US is fortunate to be self-sufficient in oil and gas production, which was a direct result of the shale revolution that took off in 2003. Because of this, the US opened the gate for the export of oil and gas in 2015. Export revenues of LNG grew exponentially over the next six years.

LNG was like a security blanket for Europe.

Russia invaded Ukraine in February 2022. Soon after, Russia cut back on gas supplies to Europe. It has been said that LNG from the US saved Europe. A connection was established between the gas-rich Permian basin in Texas, the LNG export trains along the Texas and Louisiana coasts, and export tanker shipments to Europe—important themes in this book.

One more theme is the clean-burning nature of natural gas and LNG. These burn twice as clean as oil or coal, meaning only half of the GHG such as CO_2 are emitted. The oil and gas industry has been saying for years that gas is a halfway house to net-zero.

If LNG or gas replaced all the coal, as used in coal-fired power plants, the total global fossil-caused emissions would drop from a level of 75% of all emissions to roughly 50%. This would be an enormous benefit in the race to reduce carbon emissions, because natural gas is plentiful. Now, it could be argued that a better strategy is to stop burning all fossil fuels, which would make the global 75% fall to 0%. But this is too unrealistic to happen even by 2050.

Actually, the numbers look even better than falling from 75% to 50%, because cars are switching from oil/gasoline to renewable wind and solar power, which have virtually no carbon emissions.

This chapter will emphasize how LNG is a halfway house for carbon emissions but could be called a full house for energy security.

Cheniere Energy is a big LNG exporter in Louisiana. In early 2023, the company announced it would expand its facilities in Sabine Pass (Drane 2023). It plans to boost capacity by 20 million tpy, to 50 million tpy, by adding three trains in an adjacent area, which would come online by late 2028.

Freeport LNG is one of five LNG terminals along the Gulf Coast that helped turn the US from a big gas importer to a near-the-top world exporter (Baddour and Erdenesanaa 2023). Freeport had to fix problems that led to an explosion in late 2022 before it could rejoin the market boom.

It was only a few years ago that potential investors in LNG were hesitant. One reason was fear that renewable energies would come on strong and new oil and gas assets might get started but then stranded. Instead, Russia came on strong and attacked Ukraine, and this further destabilized the oil and gas markets, which were already panicked after the pandemic.

Much of Europe's gas imports coming from Russia in 2022 were cut off by Russia. LNG was a natural replacement, and countries such as the US and Qatar and Australia stepped up to fill the gap. The export market for LNG started to explode, particularly in the US.

What Defines a Golden Age?

Pictures convey this best. Let's start with the shale revolution. Fig. 4-1 defines the golden age of shale-gas in the US, starting around 2003, when shale-gas production was initiated in several different states.

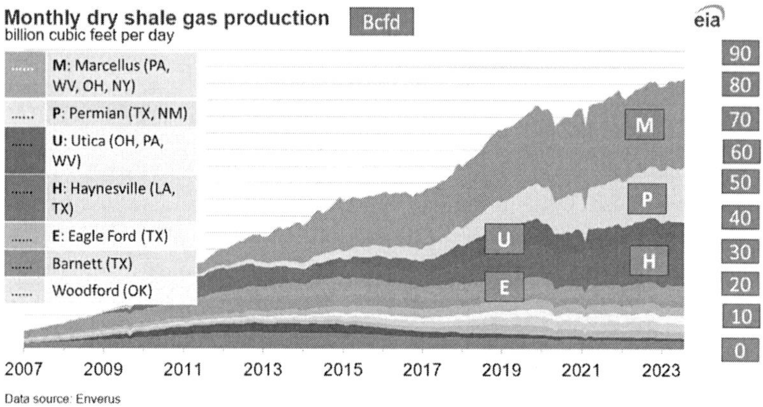

Fig. 4-1. Main contributors of shale-gas in the US through 2023 (EIA 2023a).

The Marcellus Shale is the winner and is called the *queen of shale-gas*, producing about 27 Bcfd.

The Permian is number two, at about 17 Bcfd, and its gas is associated with shale-oil production, which was booming at 6 MMbpd in late 2023. The rapid uptick in oil production began in 2011. Enfolding West Texas and Southeast New Mexico, the Permian is called the *king of shale-oil*.

LNG Boom to Europe

Out of the boom in shale-gas came the boom in LNG. It could be said that shale saved the US by making it self-sufficient in oil and gas, while shale-sourced LNG saved Europe from energy shortages caused by Russia's gas cutoff.

In the first six months of 2022, the US shipped 39 billion cubic meters (bcm) of LNG to Europe (Renshaw and Disavino 2022). This is high compared to 34 bcm for all of 2021.

The numbers confirm this (Maguire 2022). In January 2021, Europe received 45% of its total gas supplies from Russia and only 14% from LNG. By February 2022, Russia had fallen to 20%, while LNG had increased to 40%. By December 2022, Russia was at 5%, and LNG was at 53%.

Within two years, LNG had replaced all the lost gas from Russia. Most of this replacement occurred in one year, the first year of the war on Ukraine.

LNG tanker cargoes shipped to Europe shot up by 142% from about 350 in 2021 to almost 850 in 2022 (API News 2023).

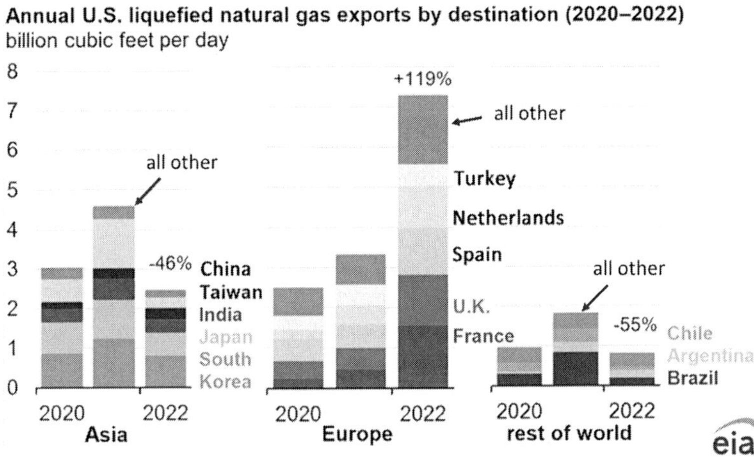

Fig. 4-2. LNG exports from US, by destination, through end of 2022 (EIA 2023b).

Southeast Asia's Place in the LNG Market

But does all the LNG that is exported from the US go to Europe?

In 2021, before the start of the Ukraine war in February 2022, Asia took the major portion of US exports, and the biggest buyers were China, Japan, and South Korea (fig. 4-2). But in a dramatic switch during 2022, Europe became the dominant destination and was taking almost 70% of the US supply—France, UK, and Spain were the top three destinations.

Meanwhile, in 2022 South Korea and Japan were the top importers in Asia (EIA 2022). China was the biggest importer in 2021 but has since fallen back because of the COVID-19 aftermath of business shutdowns in China.

With the war in Ukraine continuing, Europeans will keep on buying spot-price LNG. In the long term, however, they are committed to renewable energies such as wind and solar and grid batteries to achieve their goal of net-zero carbon emissions by 2050. They could have become cautious about signing twenty-year contracts with the US. But in 2023 the situation appeared much the same as in 2022, judging by the first three months of 2023 when overall purchases of LNG by Europe and Southeast Asia stayed in the same proportions (Russell 2023).

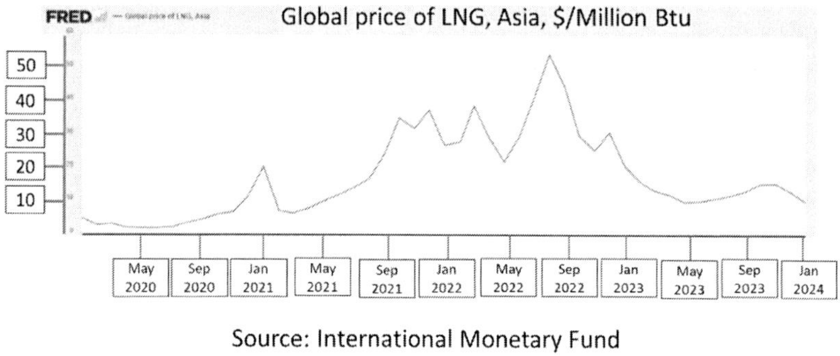

Fig. 4-3. Monthly Asian spot price for LNG in USD per MMBtu (International Monetary Fund n.d.).

The August 2022 monthly record spot price was $54/ million British thermal units (MMBtu), while the daily record was over $70/MMBtu (fig. 4-3). After this, the spot price of LNG drifted downward to $19.50/MMBtu in Q1 of 2023 (Chow 2023). The price fell to $10/MMBtu in 2023, and again in January 2024, which is as low as it had been since early 2021.

But below $20/MMBtu, LNG becomes competitive with the price of coal, and undeveloped countries such as India and Bangladesh will be attracted to LNG. Buyers would be sure to come to the LNG market if the spot price fell to $15/MMBtu.

But buyers and sellers will need to keep a wary eye on China because of its size—it might try to undercut any new deal with suppliers if its economy picks up speed. Companies owned by China signed deals with Oman and Qatar near the end of 2022. Sinopec sealed a twenty-seven-year deal with Qatar Energy in November 2022 in the largest single LNG sales and purchase agreement on record. The golden age of LNG will continue.

LNG is Like a Gold Rush for the US

After the US opened its gates in 2015 to export LNG, the shale revolution in the US became a golden age for LNG. The US has pushed aside uncertainties about stranded gas facilities and is again investing in LNG supplies. In 2022, LNG exporters completed forty-five long-term contracts to sell US gas overseas, up from fourteen in 2021 and three in 2020 (Baddour and Erdenesanaa 2023).

Fig. 4-4 shows the amount of LNG exported from the US over time. It's a strong and steady growth curve, except for the dropout during the 2020 pandemic. The export revenues surged exponentially from \$8.3 billion in Jan.–Sep. 2021 to \$35 billion in Jan.–Sep. 2022 (Maguire 2022).

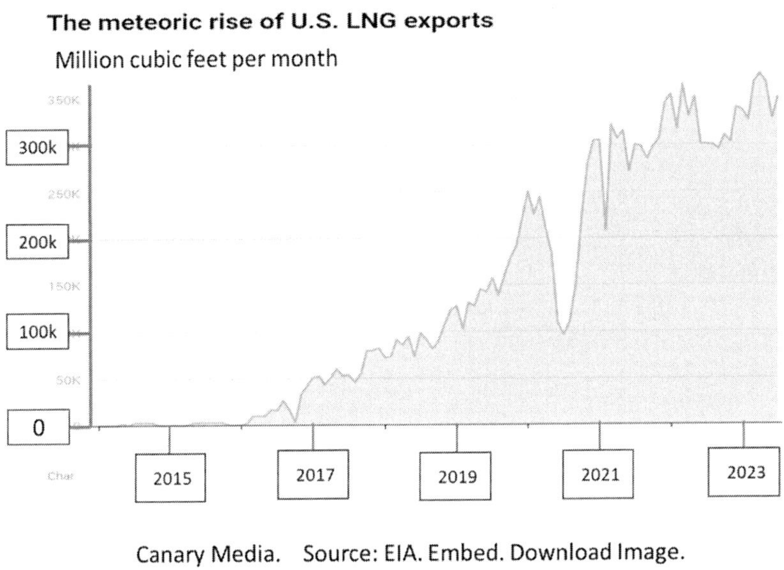

Fig. 4-4. Eye-popping growth of US exports of LNG into 2023. (Reproduced by permission from Canary Media and the US Energy Information Administration; Olano 2023.)

In October 2023 LNG exports reached their previous peak set in April of that year (fig. 4-4). Sixty percent went to Europe, while 20% went to Asia. In October 2023, Asia spot LNG prices were up 70% since the start of August due to Norway maintenance and labor protests in Australia (Williams 2023a).

The EIA projects US exports of LNG to reach 10 Tcf per year in 2050 from 4.5 Tcf per year at end of 2023 (Pitts 2023b).

But there are time delays in boosting US exports. No additional LNG export capacity will become available until later in 2024. A substantial new supply of global LNG is expected to hit the market by 2026–2027—not just from the US, but also from Qatar.

One example of this delay is NextDecade Corp., who in mid-2023 secured funds of over \$18 billion to begin an LNG export train in Texas (Liao 2023). It was the third case of this kind in 2023. The company anticipates selling LNG to customers in Asia and Europe by 2027.

The G-7 nations (Canada, France, Germany, Italy, Japan, the UK, and the US) met in mid-2023 and debated the role of natural gas versus renewable energies. They agreed to support LNG providing energy security while reaffirming the need to scale up renewables to meet the net-zero 2050 goal.

Andrew Light, US assistant energy secretary, said in a separate interview that abated natural gas should be viewed as a clean energy source (Gordon and Kumagai 2023). *Abated gas* means clean gas produced without flaring wells and without methane leaks in wells and pipelines, during transportation by road and tanker, or during the liquefying stage for LNG. But it's still not clean and green because it burns and releases CO_2 to the atmosphere—although only half the amount of that released when burning oil or coal.

The top three LNG exporters in 2022 were Australia, the US, and Qatar, just about tied in numbers (Pitts 2023a). These three were responsible for 60% of all exports. Russia came in fourth—surprisingly, in view of the war on Ukraine. Worldwide LNG trade grew by about 7% from 2021 to 2022, and the US was the main reason for the export increase—their exports increased by 15%. Europe accounted for the biggest rise in LNG imports, up by almost 70% in 2021.

The US surpassed both Australia and Qatar to become the world's top exporter of liquefied natural gas in 2023, shipping an average of 11.9 Bcf/d, up 12% compared to 2022, according to the EIA. Europe accounted for close to one half of US export volumes (EIA 2024c).

New-Build LNG Projects

Export LNG terminals are not cheap to build—about \$10 billion each (Palmer 2022e). In North America, Cheniere Energy approved a terminal expansion in Texas. In Qatar, ExxonMobil and Shell are involved with projects to grow LNG exports that cost up to \$29 billion.

Since the war in Ukraine began, about twenty import LNG terminals have been initiated. Germany has invested \$3 billion to implement four floating terminals. In China, which bought the most LNG in 2022, ten new

LNG terminals were planned to come online in 2023. China's capacity will double between 2020 and 2025.

Export volumes and wealth from LNG could potentially lead to $100 billion in new LNG developments in the US, according to Wood Mackenzie (*OGJ* Editors 2023).

Wood Mackenzie predicted that by 2027, the US will have doubled LNG production compared with Qatar and Australia, which until then will have stayed relatively constant. By 2030, the US will have doubled Qatar's increased capacity and tripled Australia's 2023 capacity, which will have stayed much the same.

This projection has included *possible* LNG new projects, and some of these are likely to be delayed or canceled—so it's surely an overestimate. Still, $100 billion cannot help but grab attention.

The head of gas and LNG asset research for Wood Mackenzie, Giles Farrer, said the following:

> Record-high prices and the need for energy security drove buyers, which included portfolio players and US producers and infrastructure companies, to seek long-term US LNG deals in 2022 and created huge contracting momentum for projects. Last year alone [2022], 65 million tpy [metric tons per year] of long-term US deals were signed, dwarfing the 18.5 million tpy we saw in 2021. (*OGJ* Editors 2023)

Glenfarne Energy Transition received final approvals in January 2024 for their new LNG train in Brownsville, Texas, that will ship about 4 million tpy of LNG (Hart Energy Staff 2024). The company beat the pause in LNG permits invoked by President Biden in that same month. Glenfarne will begin construction in 2024 and plans to start exporting LNG in 2028.

What to remember: The shale revolution saved the US oil and gas industry. And LNG from US shale saved Europe when Russia cut off gas supplies.

As always, there are headwinds. New LNG projects face mounting costs for financing and construction, delays occur in obtaining and finalizing long-term contracts, and stiff competition among developers can set back a project. Wood Mackenzie added that new LNG supply is being built because investors have gambled on Asia to reduce coal usage and Europe to continue replacing gas from Russia. The market will be subject to volatile movements in LNG price until the next uplift in supply rebalances the market and the LNG demand peaks in 2030.

Di Odoardo from Wood Mackenzie said, "The recent signing of 8 million tpy of LNG contracts from Qatar and its partners Shell, TotalEnergies, and Eni to Europe through 2053 underpins supplier confidence in the longevity of European LNG demand" (Casey 2023).

Qatar is playing a long game. QatarEnergy announced in February 2024 its plan to develop an LNG expansion of the western part of its prolific North Field—a supergiant gas field with recoverable reserves of 2,000 Tcf that blows away 281 Tcf in the Delaware field of the Permian basin, (O'Reilly 2024; Palmer 2020a).

The new project, named NFW (North Field West), will provide an additional 16 million tpy of LNG. This will boost QatarEnergy's LNG capacity to 142 million tpy. World demand of LNG is predicted to be 200 million tpy over the decade 2023–2033, with the largest demand growth from South and Southeast Asia. So QatarEnergy's new 16 million tpy will fit nicely in here.

The announcement is a timely one for QatarEnergy but not for the US, where President Biden placed a pause on permit approvals for new LNG projects in January 2024 (see chapter 17 for more on this).

Takeaways

LNG or natural gas can be viewed as a halfway house until net-zero is achieved. If gas, which burns cleaner, replaced all the coal in the world, the total global fossil-caused emissions would drop from about 75% now to roughly 50%. This would be an enormous benefit in the race to reduce carbon emissions by 2050, because gas is plentiful.

In mid-2022, the European Parliament backed a rule that classifies natural gas and LNG as *green* or *climate-friendly*, insofar as investments in these resources are concerned. This is probably a realistic reflection of the role of LNG in its golden age.

LNG has proven to be a practical boost for energy security across the world. In response, US export revenues of LNG grew exponentially through 2022 before flattening. The revenues surged from $8.3 billion in Jan.–Sep. 2021 to $35 billion in Jan.–Sep. 2022.

LNG is ideal for energy security and necessary for Europe since Russia cut the gas pipeline. Within two years, LNG replaced all the lost gas from

Russia. Most of this replacement occurred in one year, the first year of the war on Ukraine. At the end of 2023, Europe took 68% of US exports, while Asia took 19%.

But Europe may become cautious about signing twenty-year contracts with the US because of Europe's philosophy toward the energy transition. If China's economy picks up more speed, it will want to increase its purchases of LNG—which totaled over 6 million tpy from all countries in mid-2023.

Export volumes and wealth from LNG could potentially lead to $100 billion in new LNG developments in the US.

A new project named NFW (North Field West) will provide an additional 16 million tpy of LNG. This will boost QatarEnergy's LNG capacity to 142 million tpy.

CHAPTER 5

A FRESH READING OF EXXON'S PREDICTIONS OF GLOBAL WARMING AND CLIMATE CHANGE FROM FORTY YEARS AGO

After Exxon correctly interpreted global warming forty years ago, they also said some climate changes are less predictable, which brought on criticism that Exxon cast doubt on climate science.

Exxon has a checkered history in regard to climate change. Forty years ago they demonstrated strong leadership by predicting global warming from GHG emissions—and got this right. After this was a twenty-year period of quiet, and perhaps uncertainty. But in the last few years, ExxonMobil has adjusted its purpose and is bending its arc toward climate change, with strong leadership once again.

An Exxon study done in 1982 focused on global warming by increasing GHG emissions (Palmer 2020a). Remarkably, Exxon accurately predicted that by 2019, the earth would hit a CO_2 concentration of 415 parts per million (ppm) and a temperature increase from 1980 of almost 0.9°C (fig. 5-1).

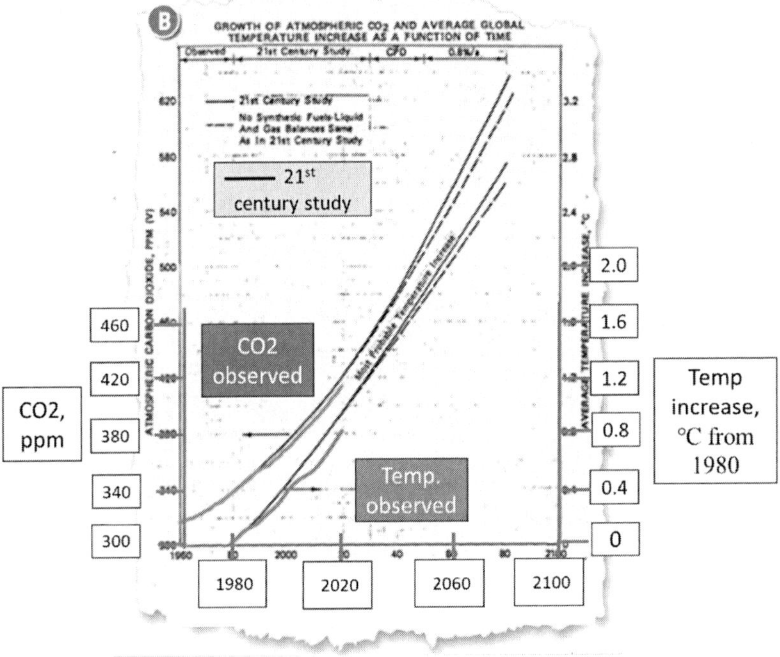

Fig. 5-1. One proprietary model of Exxon from 1982. Exxon predicted (1) GHG concentration, on the left, and (2) the effect this has on temperature, on the right. (Reproduced by permission from the American Association for the Advancement of Science; Supran et al. 2023.)

Under questioning by Rep. Alexandria Ocasio-Cortez (D-NY), former Exxon scientists disclosed this prediction in a congressional committee hearing in late 2019. Exxon climate scientist Dr. Martin Hoffert (a consultant) said "We were excellent scientists" during the hearing. In the ultimate irony, the biggest US oil and gas company predicted four decades earlier almost the exact global warming situation that occurred in 2019—and that a lot of it was caused by fossil fuel companies such as Exxon (Lavelle 2019).

But Exxon apparently did nothing about it, except to shelve the study. As reported by Inside Climate News, Dr. Hoffert described his distress at the company's newspaper ads in the 1990s contradicting the science on fossil fuel emissions' link to global warming: "What they did was wrong.

They deliberately created doubt when their internal research confirmed how serious a threat it was" (Lavelle 2019).

In 2019 the company was taken to court over this, by the state of Massachusetts. The state alleged that "the company downplayed the hazards of climate change decades ago" (Clark 2023). It appears the lawsuit is ongoing, as in 2024 many other cities and states were dealing with lawsuits against the oil and gas industry, alleging it deliberately misled the public about climate change.

Predictions by Exxon 1982–2003

A new report shows other predictions from Exxon (Supran et al. 2023). They first predicted concentration of the main GHG, CO_2, in the atmosphere in 1982. The second prediction by Exxon was related to global temperature increase (fig. 5-1). Exxon's study was updated in 2003, when a more sophisticated model predicted curve 12 in fig. 5-2 (the upper dark curve), and this matched the observed data better.

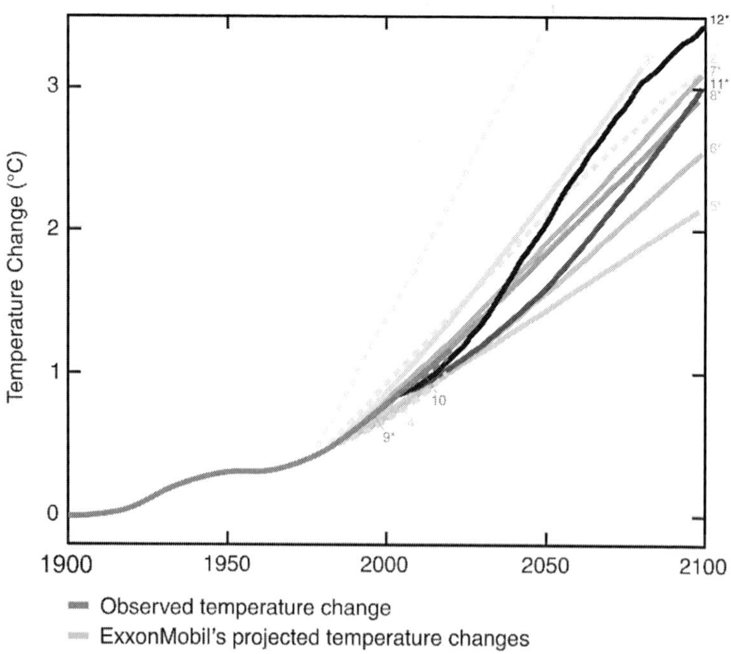

Fig. 5-2. Modeling by Exxon and third-party sources. (Reproduced by permission from the American Association for the Advancement of Science; Supran et al. 2023.)

In fig. 5-2, actual measured temperature is shown as the historical line. Modeling projections by Exxon are shown as gray or black lines. Dashed lines are projections that came from third-party sources. Solid lines get increasingly darker with the start date of the models—from light gray (1977) to dark gray (2003—top black curve labeled 12). We can assume the latter is the latest and best prediction by Exxon modelers.

These models were serious, sophisticated studies of global warming predictions that Exxon was happy to publish in peer-reviewed journals between 1983 and 1984: *Journal of the Atmospheric Sciences* and an *American Geophysical Union* monograph (Banerjee et al. 2015).

The temperature predictions out to 2020 in the 1982 and 2003 models are very well matched by actual temperature measurements (shown by the historical line). This is quite remarkable—predictions made, in one case, forty years ago and, in the second case, twenty years ago.

Exxon knew CO_2 levels were increasing over time and must have modeled a large contribution to this from the burning of fossil fuels, including their own production of oil and gas. Fossil fuels today contribute about 73% of all global GHG emissions.

A 2015 report by Inside Climate News found that Exxon scientists predicted overall warming with such certainty that it was clear the burning of fossil fuels was a major contributor (Banerjee et al. 2015).

It is clear that Exxon got global warming right, both twenty and forty years ago. But in later years it has been alleged that Exxon, in public, cast doubt on the science of climate change. Let's take a deeper look into this.

Exxon's predictions were about *global warming*, not *climate change*. Prediction of the effects of climate change is another big step and is fraught with uncertainties (Palmer 2020a). This will be discussed more in chapters 14 and 15, so we only present some results here.

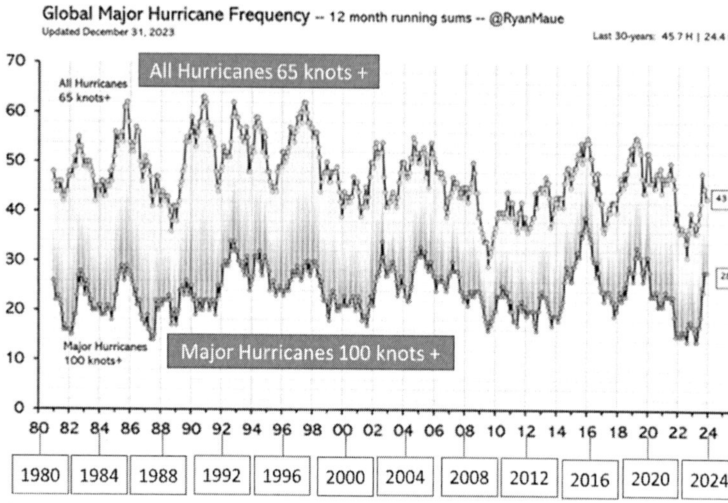

Fig. 5-3. No global trend of all hurricanes (top trace) or major hurricanes (bottom trace) in the last forty years (Maue 2024).

What is the Difference between Global Warming and Climate Change?

Some climate scientists, and many members of the press, say global warming causes more and/or worsening weather extremes. But does the data support this? Catastrophes caused by the "killer quad" of droughts, wildfires, flooding, and hurricanes have major effects on human lives, famines, migration, and stability of governments. But are the killer quad of weather extremes worsening with global warming?

On a worldwide basis, there is no long-term worsening of these extreme weather events (Koonin 2021). Fig. 5-3 shows there are not more hurricanes now than forty years ago—there is not a worsening trend. This is supported by continuous data of the other killer quad extremes over the past forty to fifty years, which will be discussed in chapter 15.

The last forty to fifty years is a period when global temperature rise has been steady and strong, rising almost 1.0°C. If the killer quad extremes were sensitive to global warming, this should show up in long-term data trends. But it doesn't (Pielke 2022; Wrightstone 2017).

But there do exist other worldwide effects, apart from the killer quad extremes. These are called *direct effects* associated with global warming.

Examples include Arctic ice melting, glaciers retreating, sea levels rising, coral bleaching, and biodiversity habitats changing. These have been serious but so far have not induced famines, or floods, or mass migrations, or government collapse on a worldwide basis. Local effects, yes, but massive catastrophes, no.

Where Does This Leave Exxon?

There is a distinction between global warming and climate change. Global warming is proven, and scientists can model it fairly well. And it's agreed that it has a human cause—largely (73%) due to burning of fossil fuels.

But predicting climate change is a different creature, rife with uncertainties with respect to catastrophes caused by droughts, wildfires, flooding, and hurricanes.

This may be where Exxon was coming from. Yes, the company was successful in predicting *global warming*. But that's not the same thing as predicting *climate change*, especially for the killer quad of weather extremes. There is a disconnect between the two. Perhaps Exxon looked at the long-term data on hurricanes from 1980 to 2000 and saw no worsening, so they waited. Perhaps they looked again from 1980 to 2010, expecting to see a worsening, but there was no worsening, so they waited. Perhaps Exxon kept waiting until the long-term data would begin to show a worsening. But Exxon saw there was no worsening that would have confirmed an important global warming prediction, so the company went quiet and waited.

Until climate models can explain why there has been no worsening of the killer quad of extreme weather events over the last forty to fifty years, Exxon would have been safe in the past by saying that climate change, and its effect on humanity, is too uncertain to be predictable.

As one example of Exxon's more recent statements of this position, chief executive Rex Tillerson said in 2013, "The facts remain there are uncertainties around the climate . . . what the principal drivers of climate change are" (Kusnetz 2023).

What to remember: Exxon predicted global warming from carbon emissions and got it right forty years ago. But climate change and global warming are different creatures

Despite statements that uncertainties make it very difficult to predict climate changes such as the killer quad, ExxonMobil has now set on a course to limit its GHG emissions. If emissions from burning its oil and gas by parties who buy its products are included, ExxonMobil's contribution to global emissions is massive because it is the biggest oil major in the US and one of the biggest in the world.

ExxonMobil is reducing methane emissions with a goal of net-zero Scope 1 and 2 GHG emissions in its Permian basin unconventional operations by 2030.

ExxonMobil has a dozen or more active CCS projects, mostly in the US. Based on this experience, they are investing \$3 billion in a new carbon company and have proposed a visionary \$100 billion joint enterprise to capture and pump CO_2 underground in the Gulf of Mexico.

ExxonMobil seems to be admitting to the public, by their actions, that climate changes may become serious and that they are embarking on these actions to reduce or bury GHG emissions. However, an independent observer might caution that ExxonMobil is not serious enough about this for the company to begin to switch from production of oil and gas to production of wind and solar or other renewable energies.

Takeaways

The predictions made forty years ago by Exxon regarding GHG and global warming were remarkably accurate. They got this right. And they agreed that global warming had a human cause—largely tied to burning of fossil fuels.

But predicting climate change is a different creature. The global data tell us that killer weather extremes were not sensitive to the global temperature rise of 1.0°C that has taken place over the past forty to fifty years.

Until climate change models can explain this, Exxon is right in saying that climate change, and its effects on humanity, are too uncertain to be predictable.

The distinction between global warming and climate change may be where Exxon was coming from. This could be important, as ExxonMobil faces many court cases across the US that have been brought against them for suppressing the dangers of global warming.

Despite their statements of uncertainty in climate change, Exxon has set on a course to limit its GHG emissions. They are investing \$3 billion in a new carbon company and have proposed a visionary $100 billion joint enterprise to capture and pump CO_2 underground in the Gulf of Mexico.

CHAPTER 6

A TRANSFORMATIVE LEAP TO CONTROL US EMISSIONS: OIL COMPANIES AND GOVERNMENT SHAKING HANDS

The costs of CCS projects under certain conditions would imply higher prices at the gas station and steeper electricity costs from coal- or gas-burning power plants.

Enhanced Oil Recovery

An insightful study of enhanced oil recovery (EOR) was made by Bryan Guzman, director of Global Business Development & Technology at Premier Corex, in late 2023. It provides an important background to what is coming—a transformative change for GHG emissions (Guzman 2023).

In the 1970s and early 1980s, Chevron led an EOR project in the Permian basin. Injected CO_2 came from natural gas processing plants and was transported over two hundred miles to the injector.

The original 45Q tax deduction for EOR, to encourage removal of CO_2 by sequestration or underground injection, was \$10 per metric ton in 2008 but was raised on a trend that would reach \$35 per ton by 2026. But in 2022 the IRA bill raised the tax credit to \$60 per ton.

From 1989 to 2016 the CCS Technologies Program at MIT reviewed CCS from a technical, economic, and political perspective. An industrial consortium was formed in 2000 and peaked at nineteen members, both heavyweights and lightweights, from the US and international countries.

In 2003 the US Department of Energy (DOE) set up, within the US, seven regional carbon sequestration partnerships (RCSPs). Their charter was to implement technology, infrastructure, and regulations to promote carbon storage.

The task in the development phase, the last phase, was to sequester at least 1 million metric tons of CO_2. In the southeast region, one of the partnerships called SECARB finally injected almost 5 million metric tons of CO_2 in an EOR/saline storage project in the Interior Salt Basin, Gulf Coast. It was called the Cranfield Project.

Two valuable outputs from these RCSPs were (1) a series of best-practice manuals for injecting CO_2 into geologic structures and (2) the National Carbon Storage Atlases (Palmer 2020b). The latter showed that the US could inject its total source of CO_2, then roughly 6 billion tpy, for twenty-three years before running out of space. Or they could store half the US source for forty-six years, or a quarter of the US source for ninety-two years. This calculation is based only on storage in old oil reservoirs, so the storage potential would be much greater if saline reservoirs were included. In other words, there is virtually no limit on potential CO_2 storage in the US.

Since 2005, US policy on carbon sequestration has been mandated through the DOE and aimed initially at emissions associated with coal-burning power plants. And the DOE did good work, as described above.

But starting in 2020, things revved up a lot in response to three bills passed by the US Congress:

- 2020: The Energy Act—DOE to establish a large-scale carbon storage program
- 2021: The Infrastructure Investment and Jobs Act (IIJA)—$110 billion for new energy and climate infrastructure
- 2022: The Inflation Reduction Act (IRA)—$369 billion for clean energy and climate investment

Energy and climate are now linked hand in hand. This period will forever be marked as a game changer for CCS in the US. A critical part of this can be seen in the latest 45Q tax credits designated by the IRA bill in 2022:

- $85 per metric ton of CO_2 permanently stored
- $60 per ton if CO_2 used in EOR projects
- $180 per ton if CO_2 obtained by DAC and permanently stored
- $130 per ton for CO_2 from DAC if utilized afterward

UIC Class VI Wells for CO_2 Injection

Permits for Underground Injection Control (UIC) Class VI wells for CO_2 injection have been mandated by the Environmental Protection Agency (EPA). But there are long delays, typically twelve months, to get a permit for UIC Class VI wells. Such delays are much less in Wyoming and North Dakota because these states have obtained special treatment called *primacy*. Wells can be permitted after three to six months instead of twelve months. Other states have applied for primacy.

At the end of 2023, Louisiana was the third state to achieve primacy, after a lengthy two-year wait on the EPA (Morris 2024). Twenty-two permits in line at the EPA will be transferred to the state. Nearly fifty CCS projects are in the works within the state. Officials anticipate a logjam of applications to the state for licenses to inject CO_2.

CarbonSAFE Program

Near the end of 2023, the DOE awarded the Projeo Corporation, part of GTI, an amount of $38 million for the Phoenix Project—Projeo's endeavor to prove the viability of injecting CO_2 into an old oil field in the Permian basin to make it a dedicated CO_2 storage region. This is one of seven projects selected for CarbonSAFE Phase III: Site Characterization and Permitting for sites that have previously done detailed studies of conditions in the subsurface.

Massive Budgets

The budgets in the IIJA and IRA bills are shown in fig. 6-1. Two columns are assignments to energy and climate, and they are massive amounts of money: $110 billion for new energy and climate infrastructure, and $369 billion for clean energy and climate investment.

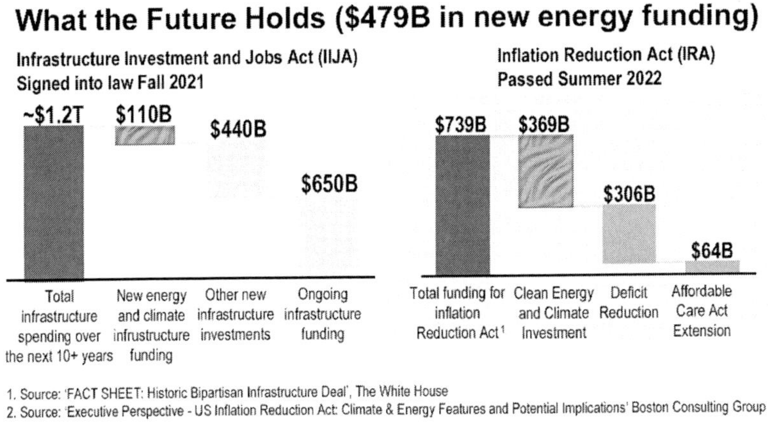

Fig. 6-1. Funding for energy and climate ventures in two US congressional bills. (Reproduced by permission from the Boston Consulting Group and Premier Corex; Guzman 2023.)

Some infrastructure funds have been awarded: $251 million to support CO_2 transport and storage—mostly large-scale commercial storage projects. Nine projects across six states have been selected, including universities, corporate hubs, and BP. But this is small potatoes compared with the next level of funding.

$2.25 billion is offered for proposals under extensions of the CarbonSAFE program. Most of this lies under two specific avenues: site characterization and permitting (up to forty separate awards) or construction (up to twenty separate awards). As an observer said, it's a bit like trying to drink from a fire hose. There is no question this government funding will kickstart a huge rush of activity in carbon storage.

A Transformative Leap in Carbon Removal by CCS

Since the 1970s, EOR has been developed and refined to sweep out the last oil residuals from old reservoirs. No great breakthroughs here. Beginning in 1988, partnerships between universities and industry highlighted advantages and problems for CO_2 storage.

Then US involvement via DOE became serious over the period of 2003–2019, with a research emphasis. Many best-practice manuals were written, including an atlas of potential storage sites across the US. The capacities of old oil fields far exceeded the CO_2 storage that will ever be needed.

Since 2008, when the 45Q tax credit was established, the US government has continued to provide funds for large-scale commercialization in the CarbonSAFE program.

Congressional legislation in 2021 and 2022 has upped the 45Q tax credits and boosted direct funding related to CCS projects by an enormous amount—by almost half a trillion dollars ($479 billion in fig. 6-1) over the next five years. This is nothing less than a transformative leap in carbon control by CCS.

The beginnings of such a leap can be seen in fig. 6-2. As of July 2023, total projects were up to 392. Total capacity rose from 244 million tpy to 360 million tpy.

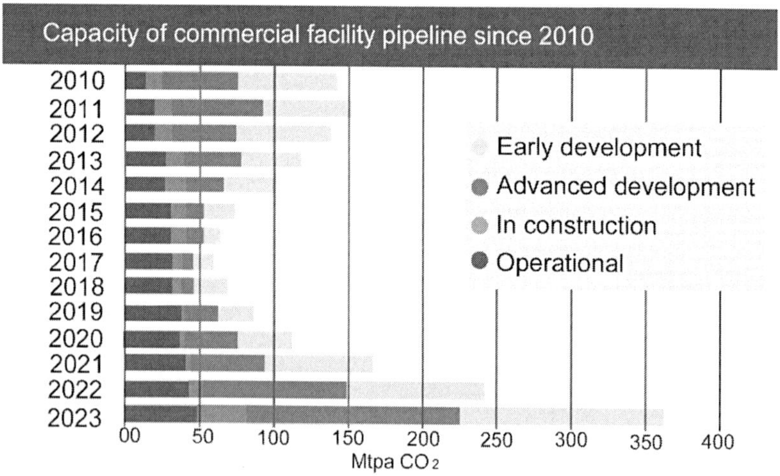

Fig. 6-2. Three hundred ninety-two total global CCS facilities were in the pipeline as of July 2023 (Daniels 2024).

The Global CCS Institute was tracking, as of mid-2023, forty-one projects in operation, twenty-six in construction, and 293 in various stages of development globally (Daniels 2024). Also, eleven new countries registered CCS facilities under development. In 2023, seventy-three new facilities entered the pipeline in the US, which dominates activity as expected.

The report offered a glowing account of progress in carbon capture worldwide, including summaries of different parts of the world such as North America, Europe, the Middle East, and Asia–Pacific. While activity is taking off everywhere, the report makes clear the world has a long way to go because current CCS is only 60 million tpy. So the plans are big, but the injection of CO_2 is yet very small.

To emphasize the transformative nature, the Inflation Reduction Act is anticipated to create more than three hundred thousand construction jobs each year plus another one hundred thousand jobs to operate these projects (St. John 2023).

Costs of Such a Leap in Carbon Removal by CCS

There is one downside of such a transformative plan. The Paris Agreement settled on carbon emissions to be net-zero by 2050, and this was signed by almost two hundred countries at the end of 2015. The general idea was that fossil fuels would gradually go away (to reduce carbon emissions) while renewable energies would increase (to replace the lost fossil fuel energies). Any leftover carbon emissions would have to be removed—for example, by underground storage of CO_2.

Fig. 6-3 shows that stopping temperatures from rising by 1.5°C–2°C by 2050, the most likely scenario, would require a global capacity increase by one hundred to two hundred times the present capacity of 40 million tpy (Abramov and Lam 2023). This would represent something like a 20% growth in CCS year over year—a scenario that's hard to fathom (Palmer 2021b). The carbon capture industry would grow as big as the present oil and gas industry by 2050, according to Rystad.

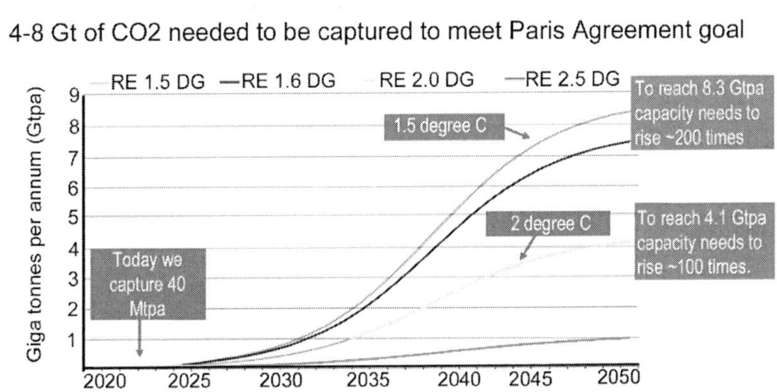

Fig. 6-3. Estimated global capacity needed for CCS to meet goals of Paris Agreement. The plot terminates at 2050 (Rystad Energy's Energy Transition Marathon, May 2023)

Two tempting hints of this eye-popping growth have already emerged. Oxy's first big DAC project in the Permian basin, to pull 500,000 tpy of CO_2 out of the atmosphere, will start commercial operations by mid-2025 (Jacobs 2023a). This is almost 0.01% of total US emissions of CO_2. Up to one hundred scaled-up facilities capturing 1 million tpy are projected by Oxy to be deployed around the US by 2035, and the CO_2 pulled out of the atmosphere by these facilities would add up to almost 1.6% of US emissions.

Second, ExxonMobil is developing a major CCS project in its Baytown refinery to inject CO_2 beneath the Gulf of Mexico. A future vision includes an expansion with other industrial partners that would produce blue hydrogen by reforming methane and dispose, by CCS, a collective 100 million tpy of CO_2 by 2040 (Woods 2023). This also is equivalent to 1.6% of US emissions.

Fig. 6-3 is based on the carbon emissions that are *left over*, meaning what's left after renewables and nature-based removal (the low-hanging fruit) have decarbonized what they can in transport and power plants.

Part of the carbon emissions left over will be from hard-to-abate sources such as cement, steel, and chemical factories, fertilizer plants, aviation fuel, and shipping fuel.

The other part will be carbon emissions from fossil fuel companies that keep on producing and burning oil, gas and coal. For this, CCS would be an escape hatch for what may become reality, as well-informed predictions of oil and gas usage range near 50% of total energy consumption by 2050 (compared with 57% now).

Other Downsides to CCS

Some oil and gas majors have diversified their portfolios by trading oil and gas operations for renewable energies—such as BP, Shell, and TotalEnergies, all European-based enterprises.

Other oil and gas majors, such as ExxonMobil and Chevron, have laid their focus on reducing carbon emissions rather than cutting back their fossil fuel production. This approach would mean, first, the world wouldn't need to build so many wind and solar projects and, second, the world would have to work harder to get rid of the continuing carbon emissions from those no-cut-back fossil fuels.

The latter would be an extra dimension to the transformative leap in carbon removal via CCS. And this would mean a much greater CCS industry. The costs of CCS projects under these conditions would be passed along to the consumer and would imply higher prices at the gas station and steeper electricity costs from coal- or gas-burning power plants (Palmer 2021b).

Second, regulations for a CCS project could be overwhelming. It's not just injection wells that have to be regulated, as described earlier, but also the technology that captures the CO_2, pipelines to transport the CO_2, and qualifying the reservoir as a safe permanent sink for the CO_2.

What to remember: A flourishing new CCS industry will not save the oil and gas industry.

Near the end of 2023, the Biden administration upped the ante with an ambitious program to roll out infrastructure for CCS and hydrogen by 2030 (Osborne 2023). Under proposed new EPA regulations, coal- and gas-fired power plants would either be shut down or deploy carbon capture or burn clean hydrogen. This implies thousands of miles of CO_2 and hydrogen pipelines would have to be built. Compliance deadlines for companies would begin in 2030 under the new rules.

This has raised an outcry among electrical power utilities and grid operators, including the Electric Reliability Council of Texas, Berkshire Hathaway, and the Edison Electric Institute, the power sector's largest trade group. The facts are there is only one power plant in existence that can capture carbon—the resurrected Petra Nova plant, outside Houston. And there are no power plants that blend hydrogen to run their turbines.

There were several comments about the reported program. Warren Buffett, CEO of Berkshire Hathaway, knows something about project success, and he wrote that the EPA timeline was "a best-case scenario—one that is unrealistic and highly unlikely to occur" (Osborne 2023). Berkshire Hathaway owns assets of power plants across the US.

On the other hand, technology has been proven for carbon capture in smokestacks and for burial of the captured CO_2. There are thirty-five carbon capture projects under development at US power plants, according to the Carbon Capture Coalition.

And, of course, the 45Q tax credit will make a huge difference, even though some calculations show it is not enough. Major corporations such as

ExxonMobil, Occidental Petroleum, and Air Liquide are investing billions of dollars to bring these technologies to scale, so their momentum and influence will count a lot.

One of the big dogs in the fight also seemed to waver on the EPA timeline:

> "We are concerned that the ability of the regulated entities to meet the proposed compliance timeframes may be difficult," wrote Matthew Crocker, senior vice president of Exxon Low Carbon Solutions. "EPA's compliance timelines should be based on realistic assumptions and in light of this uncertainty, EPA should provide off-ramps for the proposed compliance deadlines." (Osborne 2023)

Injection Drilling has Begun

In East Texas companies such as Chevron, ExxonMobil, Occidental, and BP have drilled test wells in preparation for CO_2 injection. The geology is perfect for massive storage, and pipelines are in place, which avoids the difficulties of negotiating permits for them. Wells are needed to test for injectivity and for spreading of the plume of CO_2 underground. The tax credit rise from \$50 to \$85 per injected ton has made a big difference to the cost-benefit when planning for a billion tons of CO_2, and this, ExxonMobil says, could lead to a new industry worth trillions of dollars a year. This kind of drilling and injection and storage of CO_2 will transform into a new world (Drane 2024).

Takeaways

The US could inject into old oil fields its total source of CO_2, roughly 6 billion tpy, for twenty-three years. There is practically no limit on potential CO_2 storage in the US.

Congress, in 2021 and 2022, has upped the 45Q tax credits and boosted direct funding related to CCS projects by an enormous amount—almost half a trillion dollars from 2023 to 2028. This \$479 billion has guaranteed a transformative change for CCS in the USA.

The Inflation Reduction Act is anticipated to create more than three hundred thousand construction jobs each year, plus another one hundred thousand jobs to operate these projects.

CCS is presented as an escape hatch for some oil and gas predictions, which will be absolutely necessary if oil and gas still provides around 50% of the world's total energy consumption by 2050 (compared to 57% now).

Stopping temperatures from rising by 1.5°C–2°C by 2050 would require the CCS global capacity to increase by one hundred to two hundred times the present capacity of 40 million tpy. This would represent a 20% growth in CCS year over year—a scenario that's hard to fathom. The carbon capture industry would grow as big as the present oil and gas industry by 2050.

The costs of CCS projects under these conditions would be passed along to the consumer, and this would imply higher prices at the gas station and steeper electricity costs from coal- or gas-burning power plants.

Near the end of 2023, the Biden administration rolled out an ambitious infrastructure program for CCS and hydrogen by 2030. Under proposed new EPA regulations, coal- and gas-fired power plants would either be shut down or have to deploy carbon capture or burn clean hydrogen. But this is not plausible, some qualified observers say.

CHAPTER 7

EXXONMOBIL BUYS PIONEER AND FORTIFIES ITS TRADITIONAL ENERGY INFLUENCE

The merged company ExxonMobil+Pioneer will have the largest oil and gas production in the Permian basin, one of the world's premier oil fields. The company's influence will be invaluable to other oil and gas companies.

In late 2023, ExxonMobil acquired Pioneer Natural Resources for about $60 billion, the biggest acquisition since Exxon merged with Mobil in 1999 (Mathews 2023b). This makes it the largest international oil and gas company, barring Saudi Aramco.

The merged company will have the largest oil and gas production in the Permian basin, one of the world's premier oil fields, and about 11% of the total basin production. The Permian basin is close to the Gulf Coast, where methane is converted to LNG, which has entered its own golden age for marketing overseas—a clear winner for ExxonMobil+Pioneer (Palmer 2023a).

This combination of factors is like a perfect storm in the history of modern oil and gas. But the new company's influence won't stop there—it also extends into climate change. As just one element of this, ExxonMobil has announced it will reach net-zero emissions in the Permian by 2030. Other elements are even more significant (more on this later). All these add up to a gargantuan company with enormous influence in the future of not only oil and gas but also climate change. Let's drill deeper into this unique situation.

Influence in the Oil Field

The market cap (number of shares times share price) of oil and gas companies as of June 2023 is shown in fig. 7-1. The estimated value of

ExxonMobil+Pioneer was $411 billion. This is 70% higher than Chevron, roughly 41% higher than Chevron+Hess, and more than twice as high as other oil and gas majors. ExxonMobil+Pioneer has become a giant in the oil and gas space. But it is still far behind the world's largest companies, such as Apple ($2.7 trillion) and Microsoft ($2.3 trillion).

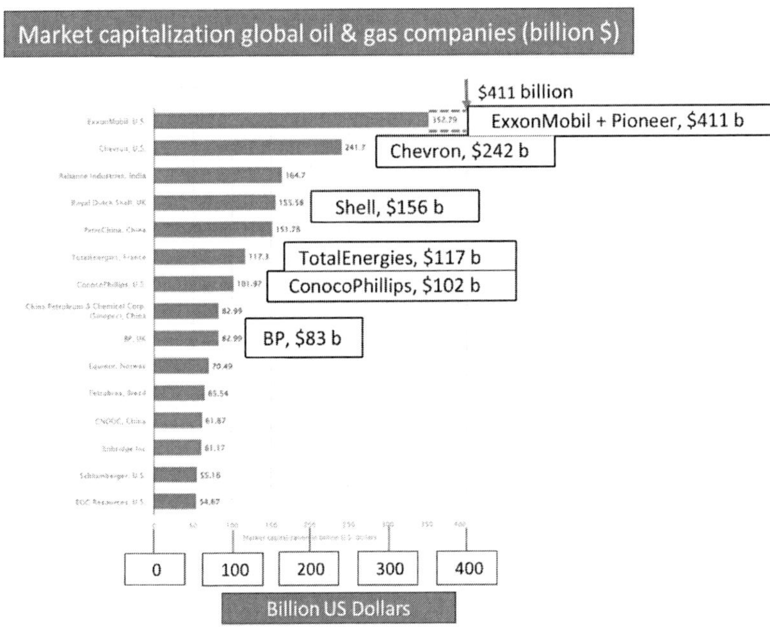

Fig. 7-1. Market capitalization of global oil and gas companies, in billions of US dollars as of June 2023 (Statista 2023b).

With the acquisition of Pioneer, ExxonMobil plans to deploy its fine-tuned Delaware basin drilling strategy into the heart of the Midland basin, where the bulk of Pioneer's leases lie (Mathews 2023c). This includes an eye-catching design where horizontal wells that are four miles long are arranged in a cubic array (Jacobs 2023b).

How much production has been added? ExxonMobil was producing 620,000 boepd, and Pioneer brought in 711,000 boepd, making a total of 1.33 millions of barrels of oil equivalent per day (MMboepd) (Mathews 2023b). This is 11% of the entire Permian basin production of about 12 MMboepd. ExxonMobil+Pioneer will be the largest shale-oil producer in the world, bypassing Chevron. Note: *oil equivalent* numbers mean oil and gas production are combined.

What to remember: The Permian basin may become a blueprint for a high-performance supply chain from upstream drilling to LNG and petrochemical products.

By 2027, ExxonMobil+Pioneer could be producing 2 MMboepd in the Permian, which would be 45% of the company's output and would be a substantial increase from the current 1.33 MMboepd. The total combined Permian reserves are estimated at 16 billion barrels of oil equivalent, which implies fifteen to twenty years of drilling inventory (Jacobs 2023b).

The ideal juxtaposition of leases between ExxonMobil and Pioneer gives ExxonMobil+Pioneer a total of 1.4 million acres. Other behemoth leaseholders in the Permian include Occidental with 2.9 million, Chevron with 2.2 million, and ConocoPhillips with over 1 million acres (Jacobs 2023b).

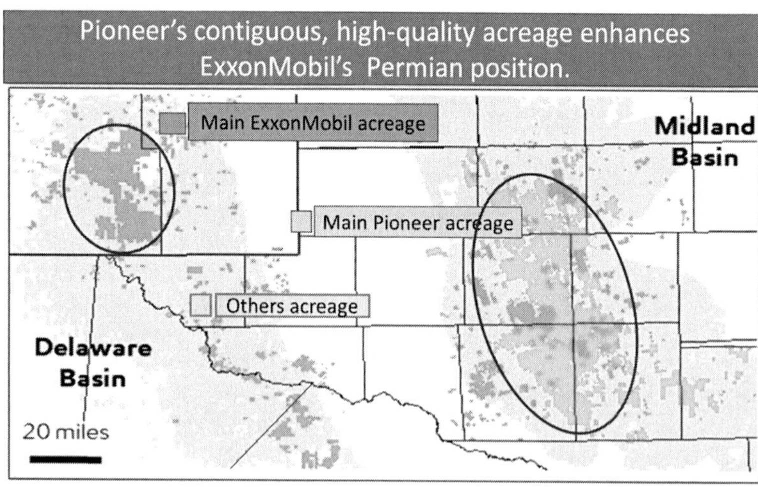

Fig. 7-2. ExxonMobil acreage in the Delaware basin fits nicely with Pioneer acreage in the Midland basin. (Photo courtesy of ExxonMobil; "ExxonMobil Announces Merger" 2023.)

The new resource base of 16 billion boe will provide everything the new company needs for a high-level operations and marketing enterprise, integrated from upstream to midstream to downstream refining and petrochemicals to exports of LNG in its golden age (Ellacott et al. 2023).

The influence of ExxonMobil+Pioneer on climate and the energy transition will also be enormous, and this will be scrutinized in chapter 8.

Takeaways

In the Permian basin, ExxonMobil+Pioneer will be producing a total of 1.33 MMboepd. This is 11% of the entire Permian basin production of about 12 MMboepd.

With its new resource base of fifteen to twenty years of drilling inventory in a time of uncertainty for oil and gas markets, the Permian merger may become a template for a productive supply chain from upstream drilling to LNG and petrochemical products. Such a high level of operations and marketing shores up energy security for the US, and overseas, in a time of uncertainty for the oil and gas industry. The ExxonMobil+Pioneer influence will be invaluable for other oil and gas companies.

CHAPTER 8

EXXONMOBIL PLUS PIONEER WILL BE *AN* INFLUENCE-SHAPER FOR OIL *AND* CLIMATE

ExxonMobil+Pioneer has set an ambitious course to limit its carbon emissions. There are three buckets to control carbon emissions: reducing carbon emissions in the field, CCS, and hydrogen generation. The company is at the forefront to repeal emissions in all these categories.

This chapter will explore ExxonMobil+Pioneer's influence in climate and the energy transition. In this sphere, it's quite remarkable what the company has accomplished in just the past few years, and it looks like this is just a start.

Influence on Climate Change in the Energy Transition

The energy transition brought on by climate change has not been closely associated with ExxonMobil until the last few years. BP had invested heavily in renewable energies, and Occidental was starting to act like a carbon-management company. But ExxonMobil?

On a closer look, history may have misjudged Exxon before and after its merger with Mobil in 1999. Exxon had a head start over other oil companies in studying global warming. In the 1970s and 1980s they predicted with stunning accuracy the rise in atmospheric CO_2 concentration up to 2020—and the accompanied rise in global temperature (Palmer 2023e).

And its human cause they also got right—largely due to burning of fossil fuels, which today contributes 73% of all global carbon emissions.

But in later years it was alleged that Exxon, in public, cast doubt on the science of *climate change*. This perception has been challenged by an alternative interpretation, as follows.

Predicting climate change is a different creature from predicting global warming. Yes, there are direct indicators such as Arctic ice melting, glaciers retreating, coral bleaching, and sea levels rising. But much more serious are weather extremes such as the killer quad of droughts, wildfires, flooding, and hurricanes. These can wreak havoc in the form of famines, flooding, migration, and government instability. But these severe effects have not worsened over the last forty to fifty years, on a global basis, while global warming has increased by close to 1°C (chapter 15 has more details) (Palmer 2023f).

Exxon may have been hesitant about forecasting weather extremes and about whether urgent action is needed to fix the problem. This may have been wrongly taken as denial of climate change because media watchers and environmental experts regularly attributed the killer weather extremes to global warming. In more recent years, other serious thinkers have also been hesitant about forecasting weather extremes, as is clear from their own careful documentation (Wrightstone 2017; Koonin 2021; Pielke 2022).

An alternative interpretation of ExxonMobil's position is that global warming is proven but that the serious weather extremes, which have not worsened over the past forty to fifty years, contradict climate alarmists. This case will undoubtedly be presented by the defense during upcoming lawsuits against ExxonMobil.

What is ExxonMobil Doing to Address Climate Change?

Despite the company's uncertainties about attributing extreme weather events to global warming, ExxonMobil has set on an ambitious course to limit its GHG carbon emissions. This makes sense because, if emissions from burning its oil and gas by parties who buy its products are included (i.e., Scope 3 emissions), ExxonMobil's contribution to global emissions will become exorbitant after its merger with Pioneer.

There are three buckets to control carbon emissions: reducing emissions in the field, CCS, and hydrogen generation. In the past few years, ExxonMobil+Pioneer has come to the forefront in repealing emissions in these three categories. Let's look closer at this surprising statement.

Reducing Emissions in the Field

ExxonMobil has developed a Permian basin road map to define its GHG emission targets. This includes a goal to reach net-zero emissions by 2030 by (1) electrifying operations using wind, solar, and natural gas; (2) expanding methane leak detection and mitigation technology; (3) eliminating routine flaring; and (4) setting up high-quality emissions offsets, which may include nature-based solutions (*2024 Advancing Climate Solutions Report* 2024).

The Permian road map will be a challenge because 2030 is not far away, and ExxonMobil's facilities have roughly doubled since the merger with Pioneer.

But the partner, Pioneer, in the past, did emphasize emissions reduction. In December 2020, Pioneer put forward a goal to lower GHG emissions by 25% and methane emissions by 40% by 2040 (Kumar 2020). In addition, they aimed to reduce well flaring emissions down to 1% of their total gas production. One percent is about the national average for flaring, but some parts of Pioneer had been above 5%.

ExxonMobil's wider goals for 2030 include reduction in upstream carbon emissions by 40%–50% (based on 2016 levels). For company-wide carbon emissions, the reduction goal is 20%–30%. For gas flaring of wells, the reduction goal is 60%–70%. For methane emissions from equipment leaks, the goal is 70%–80% (*2024 Advancing Climate Solutions Report* 2024).

Carbon Capture and Storage (CCS)

ExxonMobil has the largest number of CO_2 injection projects up and running—about a dozen, mainly in the US. The company is ready to surge out of the blocks in the race to commercialize vast numbers of CCS projects that will be needed to reach net-zero emissions by 2050.

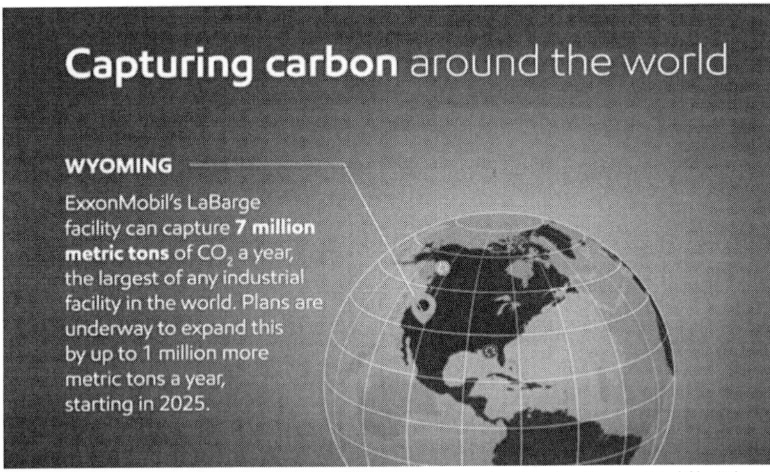

Fig. 8-1. Largest industrial capture of CO_2: LaBarge facility by ExxonMobil. (Photo courtesy of ExxonMobil.)

In 2021, ExxonMobil set up a branch company called Low Carbon Solutions to focus on reducing CO_2 emissions by scaling up CCS and new fuels such as blue hydrogen (Palmer 2023g). The two are connected: blue hydrogen is formed by the decomposition of methane (CH_4), while the biproduct CO_2 would be captured and buried via CCS in deep geological layers. There is also scope for R&D in biofuels, as well as emissions reduction.

Spending was projected to be about $2 billion in 2023, rising to $3 billion by 2025 and $5 billion by 2027 (*2023 Advancing Climate Solutions Report* 2023).

The Baytown project: ExxonMobil has advanced plans to build a 1 Bcfd blue hydrogen and ammonia production plant linked with a 7 million tpy CCS system at its refinery in Baytown, Texas (Brelsford 2023). Seven million tons of CO_2 emissions is equivalent to emissions from 1.5 million cars.

The Baytown project would be ExxonMobil's contribution to a visionary joint venture with other partners to establish a Houston carbon capture hub, valued at $100 billion. It would be built around an interdisciplinary collection of CCS and hydrogen projects.

Part of this is the company's target to dispose of 100 million tpy of CO_2 by 2040 (Woods 2023). About 7 million tpy of CO_2 will come from the Baytown refinery. The rest will come from Houston-area power plants, refineries, and petrochemical facilities that will pay to have ExxonMobil+Pioneer sequester their CO_2 emissions. This 100 million tpy is substantial: it would amount to roughly 1.6% of total current carbon emissions from the US.

The vision is a game changer, as it will reduce ExxonMobil's Baytown emissions, enable customers to reduce their emissions, and provide low-carbon hydrogen power to other Houston hard-to-abate industries.

At least eleven miscellaneous companies have expressed support for this vision (Oswald 2021). These are companies that produce electrical power, transportation fuels and oils, and everyday essentials such as medicines and plastics.

Tens of thousands of new jobs would be needed, and Houston could become the new *energy capital of the world*, not just the oil capital.

Fig. 8-2. ExxonMobil refining and petrochemical complex at Baytown, Texas. (Photo courtesy of ExxonMobil; "Baytown Operations" n.d.)

This kind of vision puts ExxonMobil+Pioneer on the big-influence stage. Their vision preceded the two US Congress bills that have since provided seed money to develop CCS: the Infrastructure Act of 2021 and the Inflation Reduction Act of 2022.

ExxonMobil's financial commitment was announced in late 2022 (*2024 Advancing Climate Solutions Report* 2024). The corporate plan for the following five years was to maintain its capital expenditures (CapEx) at $20 billion to $25 billion per year while growing investments up to $17 billion total to reduce carbon emissions. This $17 billion is still only a small fraction, 14%–17%, of total expenditures over five years.

But the Fed's elevated tax credit, 45Q, to decarbonize by using CCS, per the Inflation Reduction Act, may bring other companies running with money to pay ExxonMobil to bury their carbon emissions.

Given the massive projected growth of CCS to meet the Paris Agreement's net-zero emissions by 2050, this tax credit worth could balloon up to $100 billion. With its expertise, ExxonMobil+Pioneer is poised to cash in on this dramatic growth. But more than this—they will have the position and power to influence how this kind of energy transition develops over the crucial next decade.

What to remember: A vision for carbon emissions, CCS, and hydrogen is making *ExxonMobil+Pioneer* a powerful influence-shaper for climate projects as well as the oil field.

As an international example, ExxonMobil has signed an agreement worth $15 billion with Indonesia to invest in a petrochemical project to produce polymers, along with CCS facilities, in the country's depleted oil fields (Nangoy and Christina 2023). The CCS facilities would be the largest in Southeast Asia, and Indonesia's government is opening regulations to bring in and dispose of CO_2 from other countries. The plan would be a boost for industry growth as well as decarbonizing Indonesia and other countries in the Asia–Pacific region.

CCS Storage in Asia

To illustrate the reach of ExxonMobil, it's reported that the company is looking for storage sites in Malaysia as well as Indonesia (White et al. 2024). Developed countries like Singapore, Japan, and South Korea are releasing massive amounts of GHG but lack the geology for storage. They will need to export their GHG, and this raises a question about liability for leaks in storage sites that will exist for decades, perhaps even hundreds of years. Other major oil and gas companies are also looking at prospects to market such CCS locations for Asia exports of GHG.

Hydrogen Generation

The Baytown refinery will produce blue hydrogen plus ammonia, while the biproduct, CO_2, will be stored in a CCS project under the Gulf of Mexico. The hydrogen would be used for ExxonMobil's olefin production plant, where carbon emissions could be reduced by 30%.

The company is already talking with potential customers to purchase surplus hydrogen and ammonia volumes in a 2027–2028 time-frame.

Caution: hydrogen may not be a silver bullet for ExxonMobil+Pioneer —because it is blue hydrogen, not green hydrogen (Palmer 2022f).

One of seven hydrogen hub projects was recently awarded hefty funds (about $1 billion) from the US DOE. Called the HyVelocity Hub, it is centered along the US Gulf Coast and organized by Chevron and several private company partners, including ExxonMobil and Mitsubishi (Palmer 2023h).

ExxonMobil will have its hands on critical elements of this hub project. First, the project would be leveraging on a network of forty-eight hydrogen-production centers (the largest in the world). Second, there will be one thousand miles of dedicated hydrogen pipelines along the Louisiana and Texas coasts. Third, ten thousand permanent jobs would likely result.

Fourth are technical goals. One goal is to solve the DOE's challenge, called the Hydrogen Shot, of making 1 kg of hydrogen while emitting less than 2 kg of CO_2. Another goal is to reduce the cost of hydrogen by 80%—to $1/kg within ten years. Green hydrogen costs $3–$7/kg now, while blue hydrogen is cheaper at $1.5–$5/kg.

ExxonMobil+Pioneer will be right in the middle of this HyVelocity Hub, and its influence on R&D, hydrogen production, and miles of hydrogen pipelines will be important. Tax credits offered by the Inflation Reduction Act could enable $100 billion worth of clean hydrogen production over the coming years.

ExxonMobil as a Car Battery Supplier?

Investment in non-fossil energy is unusual for ExxonMobil. They do make hydrogen, but it's blue hydrogen—not green and clean—because it's made from methane, and the biproduct CO_2 has to be disposed of. The company had invested in making biofuel from algae but stopped its funding for this. ExxonMobil's overarching commitments have been to reduce emissions in its operations, to use CCS to bury CO_2, and to switch from natural gas to cleaner-burning hydrogen.

Now, ExxonMobil may be following in BP's footsteps by investing in versatile renewables. The company announced in late 2023 it will invest $100 million into extracting lithium for electric vehicle batteries (Frazin 2023). It has leased 120,000 acres of the lithium-rich briny Smackover Formation in southern Arkansas. As early as 2027, vertical wells at depths of 10,000 feet will produce lithium carbonate, which will be separated from produced water (Jacobs 2023c). They hope to provide lithium for batteries in a million electric vehicles by 2030.

Takeaways

ExxonMobil's hesitancy about forecasting extreme weather events from global warming, and whether urgent action is needed to fix the problem, may have been misunderstood as a denial of climate change. Other serious thinkers have also been hesitant because the killer extreme weather events have not worsened in the last forty to fifty years, as is clear from their own careful documentation.

ExxonMobil+Pioneer has set on an ambitious course to limit its carbon emissions. There are three buckets to control carbon emissions: reducing carbon emissions in the field, CCS, and hydrogen generation. The company is at the forefront to repeal emissions in these three categories.

The Permian basin road map includes a goal to reach net-zero emissions by 2030. This will be a challenge because 2030 is not many years away, and now ExxonMobil's facilities have roughly doubled, if Pioneer's facilities are included.

ExxonMobil has advanced plans to build a 1 Bcfd blue hydrogen and ammonia production plant linked with a 7 million tpy CCS system at its refinery in Baytown, Texas. The Baytown project would be ExxonMobil's contribution to a larger joint venture with other partners to establish a CCS hub in Houston, valued at $100 billion.

ExxonMobil+Pioneer will be right in the middle of the HyVelocity Hub, funded by the DOE in late 2023. Its influence on R&D, hydrogen production, and miles of hydrogen pipelines will be hugely valuable.

ExxonMobil has leased 120,000 acres of lithium-rich formations in southern Arkansas and will produce lithium from produced water as early as 2027. They hope to provide lithium for batteries in a million electric vehicles by 2030.

With its wide commitments in reducing emissions, CCS, and hydrogen fuel, ExxonMobil+Pioneer will be an indispensable influence, not just for the oil field, but also for climate projects in upcoming decades. The company will add stature to Houston becoming the *energy capital of the world*, not just the oil capital.

CHAPTER 9

OCCIDENTAL IS STICKING TO DIRECT AIR CAPTURE OF EMISSIONS, AND SO IS WARREN BUFFET

Warren Buffet may see business opportunity in CCS growth in the US. This may be a safe investment in the short term, but in the long term CCS looks to be impractical.

Since 2019 the billionaire Warren Buffet has been buying stock in Occidental Petroleum—it's now close to 4% of his holdings ("Warren Buffett Adds to Occidental Petroleum Corp Stake" 2023). Oxy's market capitalization is about $56 billion, and Buffet's holdings have reached 26% of Oxy, worth about $14.3 billion in November 2023.

Oxy's stock has seen a massive increase of about 1,000% since its IPO in 1986.

Some observers think Buffet has plans to buy out Oxy, similar to what he did with the railway conglomerate Burlington Northern Santa Fe, which he started buying in 2006 and finished in 2010.

But maybe it's just a canny investment in renewable energies. Oxy appears to be turning into a carbon-management company. They have big plans to set up CCS ventures, starting in the Permian basin, where they have operated and produced oil and gas for many years. Their new venture is DAC, where a wall of fans sucks CO_2 out of the air.

How many DAC ventures? Well, likely one hundred or more by 2035, which is a mighty big vision. Perhaps Buffet is a firm believer in the future of climate adjustments such as CCS needed to reach net-zero carbon emissions by 2050.

Or possibly Buffet is hedging for big government subsidies that will support companies such as Oxy and ExxonMobil to do CCS—by providing big tax credits like 45Q, which have jumped up to $180 per ton of CO_2 that is captured by DAC then permanently buried, a result of the Inflation Reduction Act of 2022. Such tax breaks would definitely enhance profits to be made by Oxy.

Occidental's Vision for Carbon Management

A company spin-off from Oxy, called 1PointFive, has joined with Carbon Engineering, a Canadian company, to construct a DAC facility in the Permian basin (Peters 2022).

Initially, the CO_2 removed from the air would be injected underground to produce residual oil from old oil fields. This residual oil would be labeled *carbon-neutral* because a large part (40%) of the injected CO_2 would remain stored underground.

The first scaled project, called Stratos, should come online in 2025 in West Texas. A $1.3 billion project, this will separate out CO_2 at the rate of 500,000 tpy. Other facilities are planned to separate 1,000,000 tpy. This is only 0.016% of total US emissions, but Oxy's goal of franchising a hundred such plants would remove 1.6% of US emissions.

Oxy foresees a thousand such plants across the world will be needed by 2050 to remove about 2% of current global emissions.

> **What to remember: Occidental's vision is to suck CO_2 out of the air—one hundred huge facilities in the US by 2035 and eventually one thousand franchised operations across the world.**

Oxy's second DAC has received permits for the King Ranch site in South Texas (Addison 2024). It has received a grant from the DOE of $1.2 billion from the Infrastructure Law to remove and store 1 million tpy of CO_2. But, situated in Texas not far from the industrial Gulf Coast, this DAC has the potential to remove and store 30 million tpy of CO_2.

Oxy's goal is to manufacture each of the first one hundred projects in a gigafactory. Hard to believe, but it's a giant part of their vision.

For comparison, estimates of *leftover* emissions that would need to be removed to reach net-zero by 2050 are 4–8 billion tpy (Rystad Energy's Energy Transition Marathon, May 2023). This would require six thousand of these DAC projects to get the job done.

Other DAC projects exist on a much smaller scale. Climeworks in Iceland, is capturing about 4,000 tpy. A similar plant by Heirloom Carbon Technologies is capturing 1,000 tpy in California, and using the CO_2 to make concrete, which will mean storage for hundreds of years (Plumer 2023). The first facility opened in late 2023 in Tracy, California. Heirloom has also signed an agreement with Microsoft to capture over 300,000 tons of CO_2 from the skies.

ExxonMobil's Vision for Carbon Management

ExxonMobil announced huge plans in late 2021 to capture CO_2 emissions from industrial complexes in the Houston region and store them in offshore CCS projects along the Gulf Coast (Blackmon 2022). Eleven other companies have joined with ExxonMobil in this enterprise.

As discussed earlier, the company's Low Carbon Solutions business unit will initially focus on their Baytown Refining and Petrochemical Complex. A blue hydrogen project would feed the CO_2 biproduct into a CCS operation that will be one of the biggest in the world.

This project would be a template for an integrated industry effort toward a CCS hub in Houston. ExxonMobil's goals are 50 million tpy of carbon emissions stored by 2030 and 100 million tpy by 2040. Investments would be around $100 billion.

The Problem with CCS as an Escape Hatch

Despite these game-changing visions, there is a caveat (Palmer 2021c).

If fossil energy companies continue to overproduce oil, gas, and coal (vis-à-vis the production gap), an enormous new industry for CCS will have to be created—at least as large as the present oil and gas industry according to Rystad Energy, and by some estimates, twice as large. CCS is not an escape hatch to keep oil and gas flowing.

Fossil fuel production and a new CCS industry together will be too cumbersome and expensive for energy firms to manage when compared to renewable energies. To reach net-zero by 2050, it makes more sense just to reduce the actual production of oil and gas and their emissions.

So how do we understand Warren Buffet's enthusiasm for Oxy stock? First, the push for net-zero by 2050 means many companies will want to pipeline their regular CO_2 emissions to Oxy to bury it in geological formations. The 45Q tax incentives are \$85/ton if the CO_2 is just injected and stored, or \$60/ton if the CO_2 is used to increase the flow of oil from a reservoir (EOR). Better than this, it's \$180/ton if CO_2 from DAC is stored permanently, or \$130/ton if CO_2 from DAC is used to boost oil flow. Either way, these tax incentives will make a ton of money for Oxy.

Second, perhaps Buffet is anticipating new technology will come along later, after his stock profits have been reeled in, to capture and store CO_2 and make it easier to reach net-zero by 2050 without having to build such an enormous CCS industry that will blow up the price of gasoline and electricity.

Takeaways

Oil and gas companies in the US are addressing climate change by improving the carbon footprint of their own operations, fixing methane leaks, CCS, and making hydrogen fuel.

Oxy's new venture is a DAC wall of fans that can suck CO_2 out of the air. Their vision is a hundred such facilities in the US by 2035 and eventually a thousand franchised operations across the world.

Warren Buffet's position may be that he sees business opportunity in DAC and CCS growth in the US, because oil and gas companies are preferring to control their own emissions rather than to cut back on oil and gas production.

Further, the US DOE has recently been tasked with setting up joint initiatives with the oil and gas industry to reduce carbon emissions, and this may make companies such as Oxy more profitable via government contracts that lower their direct expenses or indirectly provide tax breaks.

All the above is safe in the short term. But in the long-term view, there is a serious caveat: scale-up of CCS looks massive, expensive, and impractical to manage.

CHAPTER 10

BP IS SHAPING CLIMATE SECURITY WHILE ADJUSTING PROFIT MARGINS

The Australian Renewable Energy Hub may become one of the largest green hydrogen hubs in the world. Lots of sunshine and wind to capture for clean electricity to produce green hydrogen. And scores of markets nearby in Southeast Asia.

BP Is Integrating Rapidly

Performing while transforming is one of the catchphrases on BP's website. It refers to BP's goal to make good profits while transforming into an integrated company. Their investment in transition engines was 3% of total investments in 2019 and rose to 30% in 2022.

Some of BP's recent integrated investments in the UK and the US are compelling (Palmer 2021d). In the UK, BP's portfolio includes an offshore wind project in the Irish Sea estimated at 3 gigawatts (GW, or billion watts), installing sixteen thousand charging stations by 2030, and the visionary Teesside hydrogen project announced in March 2021.

In the US, BP is staying versatile in its new ventures. In a press release early in 2023, BP agreed to purchase 280 TravelCenters of America truck stops for $1.3 billion in cash, and they expect good returns above 15% (Angueira 2023).

BP will also install $100 million of Tesla ultrafast chargers at its gas stations and other convenience stores and truck stops around the US, tying the oil company deeper into the electric vehicle industry (Angueira 2023).

BP's Archaea Energy has brought online a renewable natural gas (RNG) plant in Medora, Indiana. It's located next to a landfill owned by Rumpke.

BP has established solar and wind projects ("Gas & Low Carbon Energy" n.d.). They have added 9 GW of solar energy and anticipate financial returns of 8%–10%. BP's goal is 50 GW of global renewables of all types by 2030. That's a stretch, but knowing BP's vision and resilience, you wouldn't want to bet against it.

To give some context, these numbers can be compared with a new coal- or gas-fired power plant, which typically generates 0.6 GW.

In September 2023, BP announced it had started construction on its 187 megawatt (MW) Peacock Solar project, located ten miles north of Corpus Christi, Texas ("BP Begins Construction" 2023). About three hundred jobs will be available during construction, and the project is expected to begin in the second half of 2024.

Peacock will sell all the power to Gulf Coast Growth Ventures, a joint venture between ExxonMobil and Saudi Basic Industries Corporation. The facility produces materials that are then used to make an assortment of clothing, containers for food, packaging, and materials used in construction.

What's interesting is the BP and ExxonMobil connection in this project—two vigorous climate influencers working side by side.

Once it's up and running, Peacock will provide a significant amount of clean energy and will avoid 250,000 tpy of carbon emissions. This is enough to power thirty-four thousand homes each year.

How much is BP planning to spend on renewables? Over the period of 2023–2030, they plan to invest $60 billion in transition growth engines (*Net Zero Ambition Progress Update* 2023). That's almost $9 billion per year. Half of this will go into convenience, bioenergy, EV charging, and later hydrogen. To be clear, transition growth includes power trading and marketing of business and convenience alongside low-carbon renewables.

Don't forget their traditional business: BP has successfully started production from the Seagull oil and gas field in the North Sea, about 140 miles east of Aberdeen. Seagull has been developed by Neptune Energy as a subsea tieback to the central processing facility that has been operated by BP for twenty-five years.

One transforming leap by BP was the industrial complex of Teesside in the UK. The vision is for Teesside to become a major hydrogen hub for aviation, shipping, and heavy trucks—all sectors where it's hard to use battery power to provide electricity ("BP Plans Major Green Hydrogen Project" 2021). But they also hope to address hard-to-abate industries such as cement and steelmaking and chemicals such as fertilizers and refinery products.

The plan is to develop a mix of blue hydrogen and green hydrogen. Blue hydrogen is the common form, comes from methane, and is cheap, but it is not clean because its biproduct is CO_2, which needs to be sequestered. Green hydrogen comes from electrolysis and is cleaner than blue hydrogen, but it is more expensive.

BP Joins with the Australian Renewable Energy Hub in Pilbara

The BP project in Western Australia is close to the Pilbara, the largest iron-ore-mining region in the world, and located close to Port Hedland. One map showed twenty-five different mines in the area. Western Australia is the largest iron ore producer and exporter in the world, with mining dominated by Rio Tinto, BHP, and Fortescue Metals.

The Australian Renewable Energy Hub (AREH) is an enterprise that plans to integrate solar and onshore wind power to generate green hydrogen and ammonia ("Renewable Energy Hub in Australia" n.d.). The AREH region is about seventy miles across. Green electricity will be delivered to the Pilbara mines, while green hydrogen and ammonia will be for domestic use and for export overseas.

Using Pilbara hydrogen and ammonia as fuel will help Australia and East Asian countries decarbonize hard-to-abate aviation, shipping, long-haul trucking, and factories making cement, steel, and chemicals. AREH is taking aim at Japan and South Korea, for example.

BP, in mid-2022, agreed to acquire a 40.5% equity stake in AREH. They will become the operator of AREH, which seems like an enormous challenge, although the company has taken on many such challenges in the past thirty years (Palmer 2021d). It's an impressive example for other oil majors that haven't yet embraced alternative energies.

The endgame is 26 GW of green power capacity, compared to 0.6 GW for a typical coal or gas power plant ("Renewable Energy Hub in Australia" n.d.). Twenty-six GW is about a third of all electricity generated by Australia in 2020. The AREH project will also generate 1.6 million tpy of green hydrogen or 9 million tpy of green ammonia.

The numbers are mind-boggling for a country that has a population of only twenty-six million and where coal power is dominant and coal exports are number one in the world.

Australia actually resisted any declaration of a net-zero emission date, such as 2050, until the year 2022. But this looks like a great leap forward for AREH, for BP, and for Australia in the pursuit of large-scale operations to reduce carbon emissions within Australia as well as East Asia.

BP Energy Outlook 2023

BP has presented its highly regarded *2023 Energy Outlook* (updated July 2023). The new perspective emphasizes the Russian war in Ukraine. This is likely to swing the pendulum away from "business as usual" and toward the Paris goal of net-zero emissions by 2050. The argument is that instability of oil and gas markets will motivate countries to develop their own renewables at a more rapid pace. BP considers three scenarios in fig. 10-1 for the future of energies and their associated emissions (Szymczak 2023). The first scenario is "business as usual," which BP calls *New Momentum*.

The second scenario is *Net-Zero*, which identifies with the goal of the Paris accords: that by 2050, carbon emissions will still be generated, but an equivalent amount will be removed or stored underground.

The third scenario is called *Accelerated*, which argues the war in Ukraine will shift the pendulum strongly from business as usual toward net-zero.

A near-term example of *Accelerated* is the act of Russia cutting off natural gas supplies to Europe in 2022, which boosted exports of LNG from countries such as the US, Qatar, and Australia. While coal and oil will decline, BP says natural gas is an X factor because of an ever-rising demand for LNG around the world (Szymczak 2023).

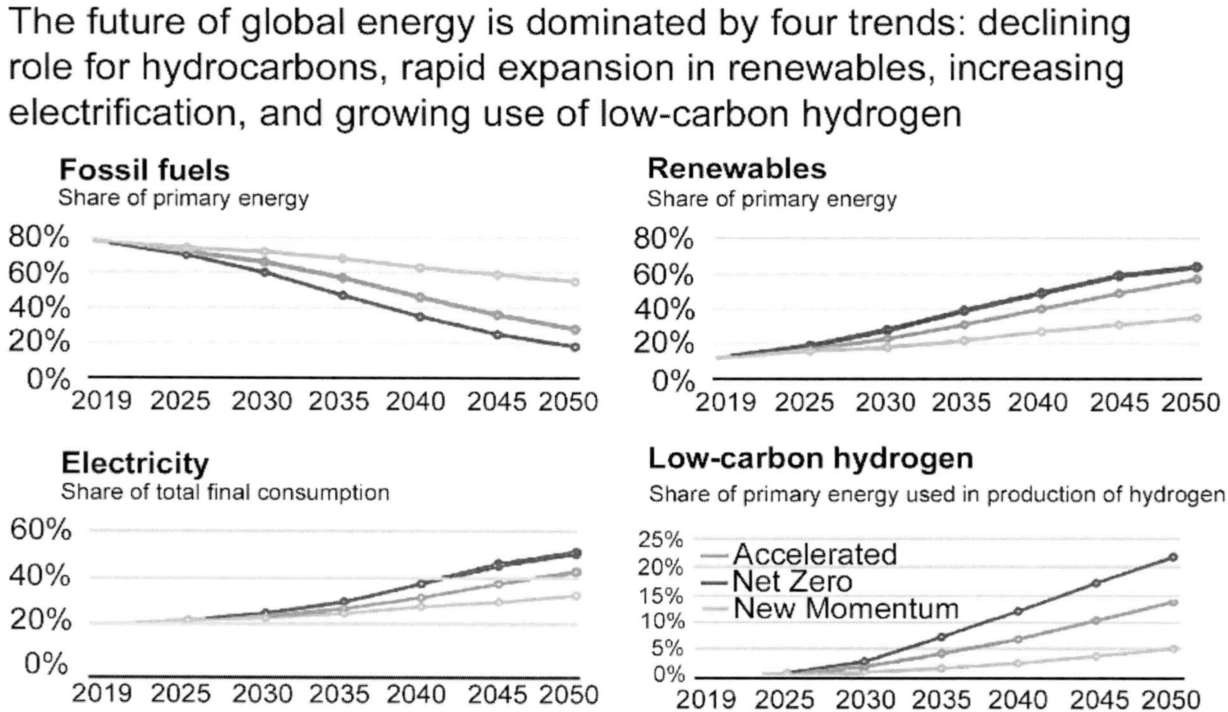

Fig. 10-1. Four BP predictions of energy through 2050. (Reproduced by permission from BP; *2023 BP Energy Outlook* 2023.)

Fossil Fuels versus Renewables

Fossil fuels fall over time for all three scenarios in fig. 10-1. They fall from about 80% of total energy consumption to 55% by 2050 under the *New Momentum* scenario. This is the smallest decline of the three scenarios, which supports the current enthusiasm of the oil and gas industry in the US, because demand and prices of oil and gas were up in 2022 and led to record profits achieved by oil and gas majors.

But due to consequences of the war in Ukraine, in the *Accelerated* pathway, fossil fuels may fall much faster—from 80% now to 28% by 2050. This is not much different from the *Net-Zero* scenario, where fossil fuels fall to 19% by 2050. Both are viewed skeptically by many in the fossil fuel industry.

In contrast to the current enthusiasm in the US oil and gas industry, BP forecasts a *drastic* reduction in fossil energies from 80% now to somewhere between 28% and 55% by 2050. Coal will fall the most, oil will fall, and natural gas may or may not fall.

Since fossil energies are responsible for about 70% of global carbon emissions now, carbon emissions would fall in concert to somewhere between 25% and 48% of global totals—a huge reduction by 2050, but not enough to get to the *Net-Zero* projection.

BP predicts that renewable energies, such as wind and solar and batteries, will rise from 12% now to 35% for the *New Momentum* or to about 60% for the *Accelerated* forecasts. In the latter case, BP has built in their expectation that the war in Ukraine and global instability of oil and gas markets will motivate countries to develop their own renewables at a more rapid pace.

BP Adjusting the Balance of Renewables

Is investment in renewables a detriment to business? This is a big question. BP has been a leader in the transition to renewables. In their new report, *2023 Energy Outlook*, BP has backed away from their goal, ostensibly because oil demand and price stayed above $80/barrel (bbl) in 2022, with a high of $128/bbl (Brent Crude price) (Edser et al. 2023). The average for 2022 was close to $100/bbl. It's tough to walk away from such a lucrative moneymaker.

BP is now expecting to produce 2 MMboepd by 2030. This is lower than 2019 levels by 25%, as compared to 40% with older plans. BP also changed its emission targets: to lower emissions by 20%–30% of throughput, as compared with a previous goal of 35%–40%.

BP's justification is here:

> BP announced it would miss its targets to reduce oil and gas production by 2030 as it said it would match investment in lower carbon projects with new investment in fossil fuels and extend the life of existing oil and gas projects. (Edser et al. 2023)

Investing in renewables is a tricky call for major oil and gas companies. European-based companies such as BP, Shell, and TotalEnergies have shifted more assets to renewables than other supermajors such as ExxonMobil and Chevron. In Europe, governments, investment houses, and stockholders have had more influence than their compatriots in the US (Palmer 2021e).

When Bernard Looney became CEO of BP in 2020, he shifted the direction of BP toward climate issues related to global warming. At the time, his economists were forecasting fossil fuel futures that would fall. If a company believes such forecasts, it's not surprising they would start shifting assets away from traditional oil and gas.

But the big question is how far to shift. BP's first goal of 40% renewables by 2030 was a hefty shift. And it's a big contrast to US-based companies that have chosen an indirect approach to climate by reducing emissions in gas flaring and methane leaks in wellheads, pipelines, and storage tanks, as well as CCS and hydrogen (Palmer 2021e).

At $28 billion, BP's profits were the lowest of the five supermajors in 2022. So it could be argued that BP's shift to renewables has hurt company profits, compared with US-based majors.

What to remember: BP diversified rapidly into an integrated company but has recently decided to slow that down.

BP is dialing back on its investment in renewables, from a goal of 40% by 2030 to a goal of 25% by 2030. This makes sense if the rationale is to capitalize on higher oil and gas prices and to maintain energy security during the Russian war on Ukraine. But it also signals the critical role that profits play in the dilemma between energy and climate security—a dilemma that faces oil and gas companies all the time.

This adjustment in renewables by BP comes at a strategic time, because in the fall of 2023, CEO Looney suddenly resigned after less than four years on the job. The reason given was not financial results but past relationships with colleagues. Looney was the mastermind who changed the goals of the company to become more integrated with renewable energies.

BP Shifts in Momentum in Early 2023

Apart from lowering the goal of renewables from 40% of business by 2030 to 25% (compared with 2019 levels), BP has changed direction in other ways. The main result is the company will retain 80% of the green electricity it produces rather than sell it to grids or utilities, where the markup is too low (only 8% returns). BP will use the power to supply EV chargers or to grow its hydrogen business. In early 2023, BP was generating 43 GW of electricity, which is substantial, and solar power made up more than half of this. Hydrogen is a big focus for BP, as described above for the Teesside and AREH projects. And in 2023, BP was the first to buy into the ultrafast Tesla EV chargers (Bousso et al. 2023).

The other renewed focus is offshore wind. BP has an established relationship with Equinor but wants to have more say in their joint projects. This includes projects in the North Sea as well as the eastern US. Equinor has more experience in offshore wind, while BP, as the larger company, has more resources to apply.

BP is not withdrawing from renewables, just rebalancing their portfolio with a close eye on investment returns. The investment returns are 15% for bioenergy/biogas and for EV charging and 10% for hydrogen business. But renewables such as wind and solar have only provided 8% returns when selling to electrical grids.

Takeaways

BP's budget for transition growth is close to $9 billion per year through 2030. Half of this will go into convenience stores, bioenergy, EV charging, and later hydrogen. Their new ventures in Pilbara, Peacock, and TravelCenters of America point to BP's versatility. Transition growth includes electrical power trading and marketing business and convenience alongside low-carbon renewables.

One early transforming leap by BP was the industrial complex of Teesside in the UK. The influential vision is for Teesside to become a major hydrogen hub for aviation, shipping, and heavy trucks as well as for hard-to-abate industries such as cement, steelmaking, and chemicals.

AREH in the Pilbara iron ore area of Australia may become one of the largest green hydrogen hubs in the world. Markets in local Australia as well as bustling cities in nearby Eastern Asia will enable AREH and help BP to reach its goal of capturing 10% of global hydrogen markets, which will testify that BP is truly integrated.

In BP's scenarios for the future, the Russian war in Ukraine is likely to swing the pendulum away from "business as usual" and toward the Paris goal of net-zero emissions by 2050.

In contrast to the current enthusiasm in the US oil and gas industry, BP forecasts a *drastic* reduction in fossil energies from 83% of total energy consumption now to somewhere between 28% and 55% by 2050. Coal will fall the most, oil will fall, and natural gas may or may not fall, because of an ever-rising demand for LNG around the world.

BP forecasts that renewable energies, such as wind and solar and batteries, will rise from 12% of total energy consumption now to between 35% and 60% by 2050 in the most likely scenarios.

But the big question for BP was, and still is, how far to shift toward renewables. BP in 2023 dialed back on their investment in renewables, from 40% by 2030 to 25%. This signals the critical role that profits play in the dilemma between energy and climate security.

CHAPTER 11

OIL AND GAS COMPANIES ARE JUMPING ON HYDROGEN, BUT IS IT OVERRATED?

Hydrogen fuel is garnering lots of press and PR about an exciting new industry, but applications are limited, it's expensive, and profits are uncertain.

How Hydrogen Fits into Oil Companies

Hydrogen is a natural fit for oil and gas companies because of their vast experience in natural gas but also because they made record profits in 2022 and can afford to take a risk on new ventures (Palmer 2022g).

Using a clean version of hydrogen would also demonstrate that oil and gas companies are trying to align with net-zero carbon emissions by 2050, consistent with the Paris Climate Accords. But not all hydrogen is clean. And there is doubt as to whether hydrogen can meet more than a niche purpose of decarbonizing hard-to-abate sectors, which include aviation/shipping fuel and the manufacture of cement, steel, and chemicals.

The production of hydrogen suits big oil and fits with grandiose oil and gas projects of the past (Palmer 2022h). The ultimate vision is to generate or collect renewable electricity, then convert it to liquid hydrogen or ammonia energy, which is portable and can be shipped around the world to provide clean green energy for hard-to-abate sectors.

The hydrogen vision is well suited to big oil's strengths: billions of funding dollars, extensive project management, and hundreds of workers to pull it off. Think of BP funding a $9 billion deepwater oil play called Mad Dog 2 in the Gulf of Mexico that was planned to produce 140,000 boepd from fourteen new wells by 2024.

Hard-to-Abate Niches

The hard-to-abate niches are aviation and shipping fuel plus the manufacture of cement, steel, and chemicals such as fertilizers and plastics. As reported by Rystad Energy in 2021, the fractions of total global CO_2 released in 2018 were as follows: aviation 2%; shipping 2%; cement 6%; metals like steel 1%; and chemicals 2%. These add up to a total of 13%.

For example, cement making uses clinker that comes by heating limestone to very high temperatures. This releases carbon trapped in the limestone, and the carbon combines with oxygen to form CO_2.

In fig. 11-1, the energy-related baseline is 40 gigatons per year (Gt/yr) for equivalent CO_2, which also includes other GHG such as methane, nitrogen oxides, and hydrofluorocarbons (HFCs).

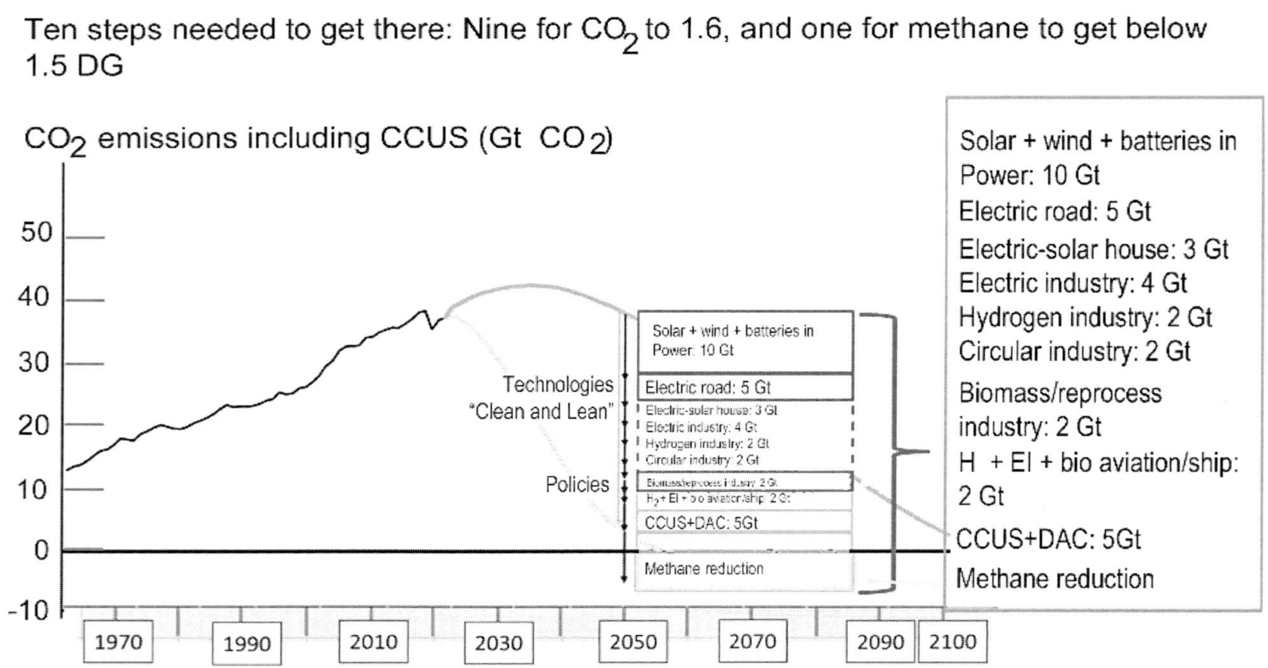

Fig. 11-1. Different steps to reduce or remove CO_2 to get from 40 Gt/yr to net-zero for the 1.6°C pathway by 2050 (Rystad Energy's Energy Transition Marathon, May 2023).

The CO_2 reductions or removals from fig. 11-1 are listed in table 11-1.

Subsector	CO_2 Emissions	Percent of 40 Gt/yr	Confidence level
Solar, wind & batteries for electricity	10 Gt	25%	Good
Electric cars, trucks, buses	5 Gt	12.5%	Good
Electric—rooftop solar	3 Gt	7.5%	Medium
Electric industry	4 Gt	10%	Medium
Hydrogen industry	2 Gt	5%	Medium
Circular industry	2 Gt	5%	Medium
Biomass/reprocess industry	2 Gt	5%	Good
H2 + Elec + bio for aviation & shipping	2 Gt	5%	Low
CCUS + DAC	5 Gt	12.5%	Low
[Methane]	10 Gt	25%	Good

Table 11-1. Different contributions in Gt/yr to reduce or remove CO_2 [and methane] to reach net-zero by 2050 for 1.6°C pathway (Rystad Energy's Energy Transition Marathon, May 2023).

The hydrogen contributions are 2 Gt + 2 Gt = 4 Gt/yr maximum, which amounts to 10% maximum of the total 40 Gt of CO_2 emissions that needs to be reduced/removed. This leaves at least 90% of the 40 Gt that has to be reduced/removed by other means—other than hydrogen. If the forecasts by Rystad in table 11-1 are about right, hydrogen produced by the oil and gas industry will remain a minor fraction of reaching net-zero by 2050.

Hydrogen Headwinds

Although many companies are active in the pursuit of hydrogen production, only 13% of projected capacity is the subject of realistic deals with potential buyers ("Hydrogen Offtake is Tiny" 2023). Roughly 50% of these deals will be hydrogen in ammonia, and the other 50% are evenly split between green and blue hydrogen.

If the oil and gas industry is planning for hydrogen to be used in the top two rows of table 11-1, this would increase its applications. After all, a few gas-fired power plants are being modified to burn hydrogen, and Japan has cars that run on hydrogen. But hydrogen is a late and relatively unproven newcomer to an energy transition that is firmly underway. It's hard to see hydrogen replacing any more than a small fraction of the solar/wind or car batteries of table 11-1. Total hydrogen usage might rise from 10% to 15% of total energy consumption, but it doesn't seem possible that hydrogen production alone would ever replace oil and gas production.

Note that CCS plus DAC will save 5 Gt/yr of CO_2 emissions from table 11-1, but this is only 12.5% of the total that needs to be reduced/removed. However, Rystad elsewhere argues that this would require an increase of over one hundred times the present amount of CO_2 saved by CCS and DAC and would mean a CCS industry that has to expand 20% year by year until 2050, which is not plausible (Rystad Energy's 2023 Energy Transition Marathon).

Fig. 10-1 in chapter 10 shows that the amount of hydrogen that will be produced by 2050 is relatively small. Only 5%–13% of primary global energy is used to produce hydrogen in the most likely scenarios. In the less likely case of reaching net-zero by 2050, the fraction would be 20%. Hydrogen is expected, by BP, to be only a minor source of energy by 2030.

Then there is the question of the cost to produce hydrogen. The Paris Agreement requires hydrogen to meet 15% of global energy demand by 2050, according to DNV (Smith 2023). But cost estimates that include $7 trillion to make hydrogen plus $530 billion for ammonia terminals plus $180 billion for hydrogen pipelines would enable hydrogen to meet only 5% of global energy demand. Getting from 5% to 15% would cost a great deal more.

What to remember: The principal use of hydrogen is for hard-to-decarbonize sectors that amount to less than 15% of global energy consumption.

Mid-2023, Wood Mackenzie published an article comparing the pros and cons of ammonia versus LNG (Farrer et al. 2023). Under a 2.5°C pathway, LNG demand is anticipated to grow over the next twenty-five years to reach 700 million tpy by 2050—a 70% increase from 2023. But under a 1.5°C scenario, the demand would reduce to 500 million tpy by 2050.

Despite uncertainties, LNG's future seems secure through 2040 because of Europe's energy security needs and Asia shifting away from coal. But after 2040 ammonia may be more profitable as the world accelerates toward net-zero by 2050.

The following are several enterprises that are promising for hydrogen production.

ExxonMobil Hydrogen

This has been spelled out in chapter 8 under the subheading Hydrogen Generation. The keys that ExxonMobil holds are, first, 1 Bcfd of blue hydrogen and ammonia in its Baytown refinery, where the company is already talking with potential customers to purchase surplus hydrogen and ammonia volumes in a 2027–2028 time-frame.

Second, the HyVelocity Hub with Chevron and others and funding from the DOE of over $1 billion. ExxonMobil+Pioneer will be right in the middle of this HyVelocity Hub, and its influence on R&D, hydrogen production, and miles of hydrogen pipelines will be important. Tax credits offered by the Inflation Reduction Act could enable $100 billion worth of clean hydrogen production over the years.

Warning: in late 2023, draft rules submitted by the Treasury Department would reduce the incentive to produce blue hydrogen from natural gas—the ExxonMobil method proposed for Baytown and its expansion to other industries. The tax credit for blue hydrogen would be moved to the lowest level, and ExxonMobil has warned that this might reduce the economics too much for these projects to move forward.

BP Hydrogen

In 2021, BP opened its hydrogen door at the industrial complex of Teesside in the UK, which contributes 5% of the nation's industrial emissions (Palmer 2021d). The vision was for Teesside to become a major hydrogen hub for aviation, shipping, and heavy trucks—all sectors where it's hard to use battery power—but also for hard-to-abate industries making cement, steel, and chemicals.

The Teesside projects of BP mesh with the goals of the UK government. Combined, HyGreen and H2Teesside could generate 1.5 GW of hydrogen production and deliver 30% of the UK government's target of 5 GW by 2030.

BP is also involved in a multinational scheme in the $36 billion AREH, an enterprise in the Pilbara iron ore region of Australia (Palmer 2022h). This will produce solar and wind energy, which are in high demand in this region of intense iron-mining featuring Rio Tinto, BHP, and Fortescue Metals. The green electricity will also be used to generate green hydrogen and green ammonia for use within Australia and for export to East Asia.

The endgame at AREH is 26 GW of green power capacity, which is about a third of all electricity that was generated by Australia in 2020. The AREH project will also generate 1.6 million tpy of green hydrogen or 9 million tpy of green ammonia.

The numbers are impressive for a country that has a population of only twenty-six million and where coal power is dominant and coal exports are number one in the world. More details are given in chapter 10.

Shell Hydrogen

Electrolysis of water for green hydrogen is more attractive now that green electricity from wind and solar is cheaper and more available. New technology known as a *solid oxide electrolyzer* promises the process will become even more cost-effective because the technology doesn't require an expensive catalyst.

In 2021 Shell launched the largest electrolyzer system in the EU—part of a plan to establish a green hydrogen network across the continent (Casey 2023).

In the same year Shell put forth a hydrogen hub concept at a Texas conference on renewable hydrogen, presaging the US government's initiative to fund up to eight hydrogen hubs in the US.

In a 1 MW demonstration project at the Shell Energy Transition Campus in Amsterdam, Shell plans to evaluate the solid oxide technology in four electrolyzer modules. When waste heat is used to generate steam, the technology should be viable when using 25%–30% less electric energy than current electrolyzers.

An interesting twist is that these electrolyzers are reversible. They can be used to produce hydrogen or, when reversed, can use hydrogen to make energy à la fuel cells.

Chevron Hydrogen

Chevron, second largest oil and gas company in the US after ExxonMobil, has produced hydrogen: blue hydrogen that reforms methane and gray hydrogen that reforms coal (Silverstein 2023). The company has retailed hydrogen via their traditional business since 2005 and now produces 1 million tpy.

Chevron has joined with the US DOE to investigate how renewable natural gas (meaning gas from landfills, etc.) can be used to generate hydrogen.

Chevron has committed $10 billion—through its division called Chevron New Business, which was initiated in late 2022—for hydrogen and renewable fuels, as well as CCS.

Chevron's larger vision is to build a green hydrogen facility in Indonesia that would make hydrogen and ammonia using renewable geothermal power of 250–400 MW. The goal is 40,000 tpy of hydrogen, or about 110 metric tons per day. But this could be scaled up to 80,000–160,000 tpy if the market is there.

TotalEnergies Hydrogen

TotalEnergies joined an Indian venture, the Adani Group, that might invest $50 billion over ten years to produce green hydrogen. In India, there is great demand for fertilizer, meaning green ammonia should have a thriving market there.

The deal was announced in June 2022, and TotalEnergies was supposed to invest $5 billion, which would amount to a 25% stake (Sharma 2023).

The Adani Group had said its initial goal was to produce 1 million tpy of green hydrogen, or about 2,700 metric tons per day, before 2030. This is an enormous number that dwarfs all other hydrogen production goals and therefore seems dubious.

But the deal was put on hold by Total Energies while they awaited results of an audit initiated by Adani due to alleged financial irregularities.

An update was provided in November of 2023 (N. R. and Crellin 2023). TotalEnergies, who already owns about 20% of Adani Green, will contribute $300 million to a new 50:50 joint venture with Adani Green Energy. The companies plan to generate just over 1 GW of solar and wind power.

Adani's eventual goal is to build one of the largest solar and wind energy parks in the world, which will produce 30 GW of renewable energy by 2030, with a promise to empower sixteen million homes. One GW is already under construction and is feeding electricity into the grid. There was no update on hydrogen production (Chetwynd 2024).

NEOM Hydrogen in Saudi Arabia

Saudi Arabia is well known as a huge producer of oil, second only to the US, and for its leadership role in the Organization of the Petroleum Exporting Countries (OPEC) and OPEC + cartels.

Also well known is the country's financial power for funding special projects. One such special project is a one-hundred-mile-long linear city called NEOM, earmarked with a cost of $500 billion (Palmer 2022i). The city is being built on a visionary approach to sustainability, clean drinking water, and climate change by demonstrating the viability of such things in the heart of an extreme desert.

The region called NEOM lies northwest of Saudi Arabia, abutting the Gulf of Aqaba and adjoining Egypt and Jordan.

What's unexpected in a country where the local prices of oil and gas are the cheapest in the world is a huge project to produce green hydrogen in NEOM ("NEOM Accelerates Progress" 2023). By 2026, "it will produce an initial 600 tonnes of green hydrogen per day, which will be available for global export, saving as much as 5 million tonnes of CO_2 emissions per year." This is equivalent to 0.08% of US annual emissions. It will be the largest production of green hydrogen in the world—on a commercial scale—with the exception of the uncertain TotalEnergies deal with Adani.

The price is estimated at $8.4 billion for this new hydrogen facility, with 73% coming from other financiers inside and outside Saudi Arabia.

Surprisingly, one goal expressed by Minister of Energy Prince Abdulaziz bin Salman Al Saud is for Saudi Arabia to be a leading exporter of hydrogen to the world. A second goal is to support a global transition to clean and sustainable energy.

The volume of clean green hydrogen produced in NEOM could save up to 5 million tpy of CO_2. This presumably refers to areas where hydrogen can replace fossil fuels in hard-to-abate applications such as aviation/shipping fuels and the cement, steel, and chemical industries. NEOM says green hydrogen could eventually make up 12% of the global energy consumption, and this 12% aligns with BP's range of 5%–13% of primary energy to produce hydrogen (see fig. 10-1 of chapter 10).

The 12% NEOM number is linked to a regional green hydrogen industry expected to be valued at up to $200 billion by 2050—an enormous number.

ABEL Energy in Australia

ABEL is an Australian developer focused on production of green hydrogen and its conversion to green methanol. SunGas Renewables, based in Texas, was chosen to supply green methanol for the $1.4 billion plant in Northern Tasmania. SunGas, a division of GTI, uses the sun's energy to convert organic wastes and their biproducts—including wood, food, animal, and human wastes—to high-quality syngas that can yield green hydrogen, RNG, methanol, liquid transportation fuels, or chemicals, with a near-zero carbon intensity.

This so-called Powerfuels Project will generate 300,000 tpy of green methanol, which is three times the current Australian consumption. It also equates to removing 540,000 tpy of CO_2. For comparison, this approximates 0.01% of current US CO_2 emissions.

A great demand is expected for methanol as a fuel for international shipping, and the hope is that Northern Tasmania will deliver ("SunGas Renewables" 2023).

Other Hydrogen Enterprises

A Denmark company called Copenhagen Infrastructure (CIP) has plans to build energy islands in the North Sea, the Baltic Sea, and in Southeast Asia. The key is to link offshore wind turbines with green hydrogen production. There are useful advantages: lower electrical power transmission costs and easier integration of wind energy into global energy systems. The founding group, Copenhagen Energy Islands, has Nordic, European, and North American investors. They anticipate $163 billion in investments in coming decades. ("Denmark to Build Hydrogen Islands" 2024). Note: the Danish government had previous plans for an energy island, but this was postponed in early 2023 due to high costs.

A Swedish project by a company called H2 Green Steel, has secured financing from many investors, including Microsoft Climate Innovation Fund.

The company, founded in 2020, is constructing a factory in Boden, Sweden. It will combine green hydrogen and green iron production, and plans to produce steel with only 5% of the carbon emissions that a coke-fired blast furnace would produce. Half of the production of 2.5 million tpy of near-zero steel has already been sold in five-year contracts ("H2 Green Steel" 2024).

Last, at the CES meeting in Las Vegas in January 2024, Hyundai presented its process to convert waste to hydrogen. The company is selling cars powered by hydrogen and are looking at hydrogen for heavy transport applications. Truckmaker Nikola demonstrated a hybrid made up of hydrogen and electric power (Kahn 2024).

Is Hydrogen Overrated?

Despite widespread interest, hydrogen has only a small investment so far. Lots of press and PR about an exciting new industry, but annual investments of a few billion dollars (max) per oil and gas major combine to less than a tenth of the global greenfield CapEx of $200 billion earmarked for 2023 (Palmer 2023i).

But the oil and gas industry can portray their hydrogen investments as a serious contribution to meeting the Paris climate goals.

On the other hand, while the oil and gas industry currently supplies about 57% of the world's energy, it's also responsible for roughly 50% of global carbon emissions, from burning oil and gas (Palmer 2020a). So even if the oil and gas industry were to provide hydrogen for all of the hard-to-abate sectors (defined earlier in this chapter) that would offset only about 13% of the total global CO_2 emissions. The remaining oil- and gas-caused emissions that are not offset amount to 37% that the industry is still responsible for.

Hydrogen might remove some CO_2 emissions from power plants, transport, regular industrial applications, and home or office heating because it is a cleaner alternate fuel, but the numbers now are small and likely to be of little significance by 2050. It's a case of too little, too late.

The remaining 37% of global emissions is the responsibility of the oil and gas industry. The industry can chip away at methane leaks, find sparse applications of hydrogen, and drill into geological layers to bury CO_2, but in the end oil and gas production will have to fall if the world will ever reach net-zero GHG emissions. Hydrogen will not be able to save oil and gas.

The following is an example from 2024 that illustrates, sadly, how rising costs have dampened investments in green hydrogen projects in Europe.

> The Innovation Fund is one of the "must-succeed" programs to ensure that new technologies can quickly play a major role in bringing down EU emissions, according to Marcus Ferdinand, chief analytics officer at Oslo-based research firm Veyt. . .
>
> One was a plan by a division of German utility Uniper SE to produce green hydrogen at a site on the outskirts of Rotterdam. In many ways it's an ideal location, close to major industrial users and right on the coast, giving it easy access to the growing fleet of offshore wind farms in the Dutch North Sea.
>
> But soaring costs in recent years for electricity, labor and financing have made green hydrogen even pricier, making it difficult to attract potential customers.
>
> "It is too expensive at the moment," said Dyonne Rietveld, managing director for Uniper in the Netherlands. "Interest rates and the costs of grid-connection fees and the risk profile of power purchase agreements are killing investment decisions." (Mathis and Ainger 2024)

Takeaways

Large and small companies are jumping on hydrogen, including ExxonMobil, BP, Shell, Chevron, TotalEnergies, NEOM in Saudi Arabia, and ABEL in Australia.

But there are serious headwinds. Reputable projections show hydrogen usage would likely rise from 10% to only 15% of total energy consumption. And one cost estimate above $7 trillion would be what it takes to make enough hydrogen for only 5% of total energy consumption. Getting from 5% to 15% would cost a great deal more.

Broad market application of hydrogen—beyond a niche market of aviation/shipping fuel and the manufacture of cement, steel, and chemicals (fertilizers and plastics)—seems very uncertain. It's unlikely that hydrogen production would replace significant oil and gas usage in cars and gas-fired power plants.

CHAPTER 12

THREE REASONS WHY OIL AND GAS SHOULD KEEP AN EYE ON THE BATTERY TRANSITION

Surging batteries are a proxy for displacing gasoline in cars and trucks and for displacing coal and natural gas in power plants.

Electricity generation uses the most energy in the US, followed by transportation, then the industrial sector. But climate goals and mandates may force oil and gas companies to pivot more toward addressing climate change. The adoption of renewable energies by government mandate, bankers' priorities, or investors' funding power may seriously change the future of oil and gas production. One of the keys to adoption of renewable energies is batteries, and their deployment is changing rapidly.

Can batteries pose a threat to traditional energies such as oil and gas? Let's just say the oil and gas industry would be wise to keep an eye on battery growth—for three reasons:

1. Small batteries provide electricity in EV cars and trucks, which will displace some gasoline and diesel, and therefore oil.
2. Home batteries store electricity from rooftop solar, which will displace some coal and natural gas that fuel power plants.
3. Grid batteries provide electricity for the grid, which will also displace some coal and natural gas in power plants.

A calculation can be made for the future of oil and gas production based on two of President Biden's goals regarding (1) new sales of EVs by 2030 and (2) how much of the US total electricity will be provided by wind and solar energies by 2030. A simple zero-sum equation will provide insights into the potential decline of oil and gas production in the US by 2030. The result is a significant decline.

EV Growth Will Affect Oil Production

The numbers below are for 2022 and they come from the EIA (2023c).

Fig. 12-1 shows that oil usage in US transportation was 24.8 quads in 2022 (1 quad is 10^{15} Btu). President Biden's goal is that new sales of EVs will be 50% of all sales by 2030. If new sales are roughly 10% of total car numbers, then 5% of total car numbers will be new EVs in 2030. Note that pure-electric and plug-in hybrids were 9% of car sales in 2023 (see later).

We can extrapolate EV growth backward by postulating 4.5% of total cars will be new EVs in 2029, and 4.0% of total cars will be new EVs in 2028, etc., all the way down to 0% of total cars will be new EVs in 2020. Summing these up tells us that 27.5% of all cars will be EVs by 2030. This is only an estimate, until a better analysis using yearly numbers is available. Let's say 25% of all cars will be EVs by 2030, and these cars are not running on gasoline.

A 25% decline in gasoline cars from today would take oil usage in fig. 12-1 from 24.8 to 18.6 quads—a 6.2 quad drop. Since total petroleum usage is 35.8 quads in fig. 12-1, this 6.2-quad decline implies a 17% decline in consumption of oil in the US by 2030.

If supply follows demand, then a 17% decline in crude oil production would be expected by 2030—almost a fifth of oil production declining by 2030. This would be a sizable hit to oil production in the US.

There is a caveat: the potential demand in the US may drop 17%, but crude oil sales abroad to places such as Southeast Asia might replace the demand and keep the supply up in the US.

Postscript: The EPA has a new rule, as of March 2024, that lowers tailpipe emissions, and it means that new sales of EVs will be 67% of all sales by 2032. So this new rule anticipates the decline in crude oil production would be significantly more than 17% by 2032.

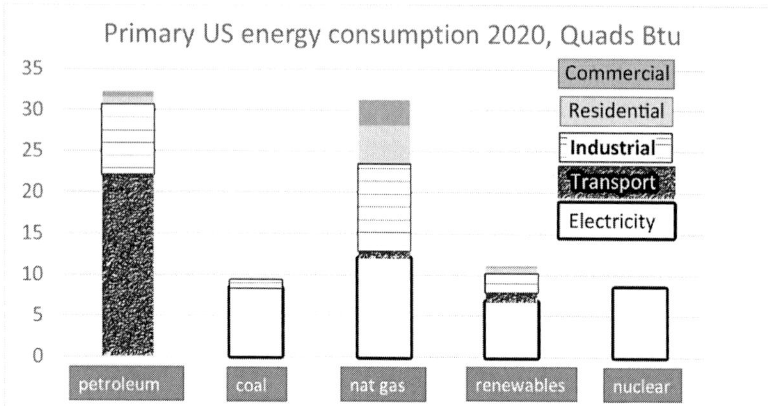

Fig. 12-1. US energy consumption by source and sector. (Figure by author with data from the EIA; EIA 2023c.)

Replacing Power Plants by Solar and Wind Will Affect Gas Production

President Biden's goal is for renewables to provide 80% of the US's total electricity by 2030.

As seen in fig. 12-1, total US electricity consumption in 2020 was 37.8 quads. Eighty percent of this is 30.2 quads. Current electricity from renewables (7.9 quads) and nuclear (8.0 quads) is 15.9 quads, which is 42% of the total. To go up from 15.9 quads to 30.2 quads would require 14.3 quads of coal or natural gas to be displaced by renewables.

Let's assume half of the coal power plants have been shut down by 2030, which is 4.9 quads. To then reach 14.3 quads would mean shutting down 9.4 quads of gas plants, which amounts to 28% of the 33.4 total quads of natural gas shown in fig. 12-1.

This ought to be a concern for the oil and gas industry, because if supply follows demand, production of natural gas in the US would fall by 28% by 2030.

The fall would be even more than 28% if less than half of the coal-fired power plants were turned off by 2030.

However, the impact of the fall would be less if surplus US production continued to be routed to Europe and Southeast Asia via LNG tankers. In mid-2023, the EIA reported a 43% increase, or 35 Bcfd, in US gas demand between 2021 and 2022—driven by LNG exports (EIA 2023d). Almost half of this increase originated from LNG in Texas and Louisiana, where gas demand shot up by 116%.

The numbers estimated in this section are only approximate. But they do warrant that the oil and gas industry keep a watchful eye on year-by-year growth in batteries, because EV batteries reflect EV growth and, second, home and grid batteries reflect renewable electricity. Home batteries store rooftop solar energy, and grid batteries store energy from wind and solar farms. It's important to understand how these batteries are key to the energy transition.

EV Batteries

President Biden's two goals mentioned in the previous section depend on batteries, which are now recognized as a key to success. Because of climate change and the energy transition, the world is entering a kind of battery revolution.

EV batteries are small batteries that provide power for electric cars and trucks. The technology is advanced and is the foundation for the rapid, exponential growth of EVs across the world (Palmer 2022j). Growth has begun in the US. In California, 25% of all new car sales were electric in 2023—nearly three times the national average (Mihalascu 2023). Washington State is second at 18%, and Oregon is third at 17%. But averaged across the US, new passenger EVs were only 7.5% in 2022. That number rose to 9% in the first half of 2023. BloombergNEF forecasts 23% in 2025 and a leap to 51% by 2030.

In the US, Tesla has led the way and even has twelve thousand charging stations built along highways. Called Tesla Superchargers, these fast and dependable battery chargers have been in the works since 2012.

Surprisingly, Ford announced it will join the Tesla network by installing adapters in 2024 and proper car connections in 2025 (EIA 2023d). GM, Rivian, and Volvo have also joined the Tesla network. Without a doubt, this will accelerate the transition to EVs across the US.

The oil and gas industry needs to keep an eye on EVs and their batteries—they are a proxy for displacing gasoline cars and declining oil production.

Gigafactories

Elon Musk's vision for Tesla EVs included the gigafactories that make them. Tesla has six gigafactories. There are four in the US—Fremont, California; Austin, Texas; Buffalo, New York; and Sparks, Nevada. Plus there are two overseas: one in Berlin, Germany, and one in Shanghai, China (Rapier 2023).

The Nevada gigafactory produces electric motors as well as batteries. Tesla announced an investment of $3.6 billion in 2023, which will be used to produce the Tesla Semi, a fully electric semitrailer that will cost $250,000 and is even now being delivered to customers.

Musk said his goal is to make twenty million electric cars each year by 2030, and this would require building ten to twelve more gigafactories.

Tesla has planned a gigafactory in Mexico, and it's not cheap. At a cost of $10 billion, this would be the most expensive one yet.

Other automakers are following in the steps of Tesla. A $9.2 billion loan has been made to Ford by the US DOE to build not one but three factories that make batteries.

Potential issues for EV batteries are availability and cost of battery metals such as lithium, cobalt, and nickel (Palmer 2022j).

Global View of EVs and Batteries

The calculations in this chapter so far are US-based. But Rystad Energy has done their own calculations for the global case, and these are shown in fig. 12-2. Three different climate scenarios are differentiated for the period 2025–2035, due to the acceleration of EVs which will displace oil consumption. Rystad argues that commercial vehicles will move the shift away from oil consumption even faster than standard cars.

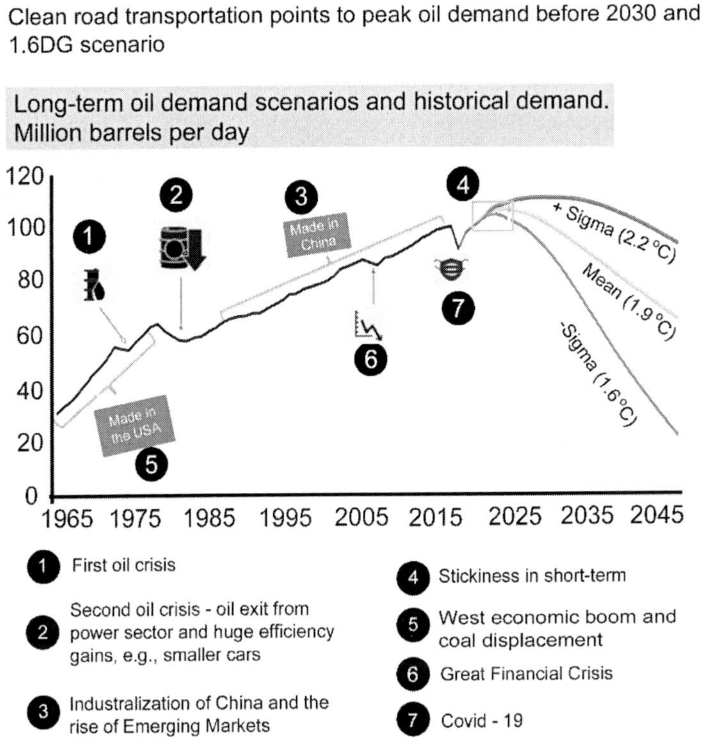

Fig. 12-2. Effect of EVs in road transportation on global oil demand (Rystad Energy's 2023 Energy Transition Marathon, May 2023).

The considerations that went into the three prediction curves of fig. 12-2 are:

- Combustion vehicle sales declining—reduces oil consumption.
- Electric vehicle sales increasing—displacing oil consumption.
- Commercial vehicles: electric drive trains gaining traction. Heavy-duty trucks: electric most likely, but could be hydrogen or biogas.
- Commercial vehicles can impact oil consumption much faster due to cost focus and short lifetime.
- Peak oil demand likely before 2030 on a 1.6°C scenario.

The results are twofold: first, peak oil may occur before 2030, shown by the 1.6°C and 1.9°C pathways on the right side of the figure. Second, global warming may reach net-zero on a climate pathway of 1.6°C—see green line. Rystad's optimism is discussed further in chapter 16.

The import of all this is that the US and the world are confronting a serious decline in oil production because of an increase in EVs and heavy-duty trucks on the road, which has started now, particularly in Norway and China.

What to remember: The US and the world will face a decline in oil production because of a growing number of electric vehicles on the road.

The uptake of EVs on the road in the US slowed in late 2023 and early 2024, but for 2023 the stats are: pure-electric vehicles increased to 7% of car sales, hybrids rose to almost 8%, and plug-in hybrids grew to 2%. Altogether, pure-electric and hybrids increased to over 16% of car sales, while internal combustion cars dropped to 84%. Hybrids still burn gasoline, so they reduce but do not eliminate carbon emissions. Note that the transport sector in the US accounts for 25% of GHG emissions (McCarthy 2024).

The Biden administration, via a new EPA rule in late March 2024, will reduce pollution from the nation's cars and light trucks by imposing new tailpipe emission limits. This will compel automakers to boost sales by 2032 of battery-electric and plug-in hybrid models to 67% of new car sales, but plug-in hybrids can make up to 13% of this (Marshall 2024; Dloughy 2024).

In-Home Batteries

In-home batteries serve two purposes. First, to store rooftop solar in the daytime so an EV can be charged overnight in the garage. Second, to offset electricity prices imposed by the grid.

Australia has some startling statistics. One of three homes in Australia has rooftop solar. The uptake there has been much greater than in the US. In 2021 there were as many rooftop solar systems in Australia as in the entire US, where the population is fourteen times larger.

One in seven of these homes in Australia has a home battery. Home battery storage now totals about 2 GW. One purchaser said the payback time was ten years, a longish time that means widespread adoption will be slower. But Australia appears to be out in front in this transition.

Australian data suggests home batteries, in association with rooftop solar panels, can reduce utility costs from the grid by a third, or 33%, on average (Waterworth 2023). In Australia in 2022, almost fifty thousand home batteries were installed, a 55% increase from 2021.

One observer said that over twelve years he'd saved about AU$20,000 (US$15,000) by installing rooftop solar and a home battery. Now that he has an EV to charge, he's installed a separate but similar system for the EV.

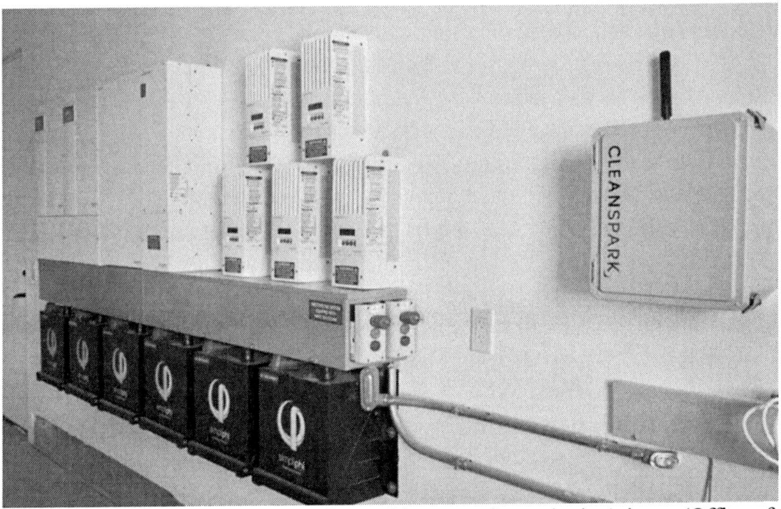

Fig. 12-3. Battery system connected to residence rooftop solar in Arizona (Office of Energy Efficiency & Renewable Energy 2021).

In Europe, batteries are growing too. Energy storage for home and utility batteries surged 64% to $8.4 billion in 2023, largely because energy prices zoomed up after Russia attacked Ukraine (Roston 2024).

Grid Batteries

Grid-scale batteries, also called *big batteries*, are needed for an electricity grid that is sourced by solar and wind power. In the US in 2022, grid-scale batteries added capacity of 4 GW, almost as much as in all previous years combined (4.7 GW) (Storrow 2023).

These batteries are needed for electricity storage to compensate for daily fluctuations of sunlight or wind power or for other emergency loss of power. Batteries are an essential part of the transition to renewables.

The obvious example of daily fluctuations is solar power. A battery that can store solar power during the day and discharge it for four hours during early evening is a big asset in the summer, when homes pull on their air conditioners. California was able to draw on this system in the summer heat of 2022, and this may have helped them avoid serious blackouts.

In the US, California and Texas, which are both strong sunshine states, lead the charge in new grid batteries. In 2022, California brought online 60% of big batteries, while Texas brought on 33%. These two states are numbers one and two in solar markets in the US.

Most of the grid batteries are of the lithium-ion type and are the same kind of batteries as used in EVs. The technology growth that was centered around EV batteries spilled over and led to lower costs of grid batteries.

Big batteries are booming. Tax credits in the Inflation Reduction Act of 2022 will support this development. The biggest US battery, built in 2023, is 750 MW/3,000 megawatt hours (MWh), in Moss Landing, California, and it is claimed to be the largest in the world (Lewis 2023).

Note: big batteries are often listed as, for example, 100MW/200MWh. The 200MWh means the battery, if running fully at 100MW, will supply power for two hours. Or if running at half power of 50MW, it will supply power for four hours.

LG Energy, a company out of South Korea, has plans to build ten grid-scale batteries in the US in 2024, totaling a capacity of 10 GWh (Johnson 2024). It's a huge expansion that is equivalent to supplying 10 GW for one hour, or 2 GW for five hours. The company builds lithium-ion batteries across the world and has announced a special factory in Arizona to build grid-scale as well as EV batteries.

Texas will beat out California in new battery deployment in 2024. Texas will add 6.4 GW, compared with 5.2 GW in California. This 6.4 GW addition will make the Texas total more than double the 5.6 GW that Texas installed in 2023. It's also more than what was installed in the entire US in 2023. Two reasons for the uplift are that land in Texas is cheaper and regulations are less strict than in California (Spector 2024).

Current development plans in the US would reach 30 GW by year's end 2024. Surprisingly, this capacity would exceed petroleum liquids or geothermal or gas from landfill waste (EIA 2024b).

A Midsize Battery Option

The governor of New York State, Kathy Hochul, set a goal in 2022 for the state's electricity to be 70% carbon-free by 2030 and to install batteries that could supply 6,000 MW when needed. A start-up company called NineDot Energy has built a midsize battery—in between home batteries and grid-scale batteries (St. John 2024a). The first such battery was built in 2022 in a parking lot in the Bronx. Its capacity is 3 MW/12 MWh, which means it can supply 3 MW for four hours. Four Tesla megapacks store the power during the late nighttime and release it for peak loads during the day, or when regular power is shut down.

Takeaways

President Biden's goal is that new sales of EVs will be 50% of all sales by 2030. This suggests a 25% decline in gasoline cars by 2030 and implies a 17% drop in crude oil production by 2030. This is based on a zero-sum equation in the US and may be upset if oil sales increase overseas.

If renewables provide 80% of total electricity by 2030, per Biden's goal, then production of natural gas in the US would fall by 28% by 2030. Again, this result may be distorted by a booming LNG market overseas.

Surging batteries are a proxy for displacing gasoline in cars and trucks and for displacing coal and natural gas in power plants. The numbers warrant that the oil and gas industry keep a watchful eye on growth in EV batteries, home batteries for rooftop solar, and grid batteries that store energy from wind and solar farms. Each of these can displace oil and gas consumption.

Other news that may impact oil and gas: Ford said it will join the Tesla network of Superchargers in two steps in 2024 and 2025. Rivian and Volvo have also joined. This should accelerate the transition to EVs across the US.

A new EPA rule that governs tailpipe emissions will also accelerate the shift to EVs and plug-in hybrid vehicles in the US.

Elon Musk said his goal is to make twenty million electric cars each year by 2030, and this would require building ten to twelve more gigafactories.

A $9.2 billion loan has been made to Ford by the US DOE to build not one but three new battery factories.

A midsize "community" battery is another battery option with promise that is now being tested in New York.

CHAPTER 13

BATTERY DEPLOYMENT AND SUCCESS IN THE TRANSITION DOWN UNDER

Seventy-two percent of South Australia's electricity comes from renewable sources, which was achieved over sixteen years. On 180 separate days in 2021, renewables met 100% of electricity demand.

Battery Deployment in Australia

Australia is right up there with the US in its fossil fuels. It's the number two coal exporter in the world (US is number four). It's about equal as the number one exporter of LNG with US and Qatar (Statista 2022). Australia is close to being self-sufficient in fossil energy, while the US is self-sufficient. These are surprising achievements because the population of Australia is only 26 million, compared with 334 million in the US.

So what can the world learn from little ol' Australia regarding their transition from fossil to renewable energies? It turns out Australia is pushing hard to bend the arc of its climate history, and some successes down there are truly remarkable.

Australia is wading into renewables, but it varies by state. Tasmania (99%), South Australia (72%), and Western Australia (31%) are states most committed to renewables. Tasmania is mainly hydroelectric, while South Australia is dominated by wind farms and big batteries.

As in the US, the uptake of EVs has been slow compared with other countries such as Norway and China. So the effect this has had on displacing gasoline cars and crude oil production is not significant—yet. But rooftop solar, including in-home batteries, has become a very big deal in Australia. This means less use of grid power, which can displace coal-fired power plants and quite possibly gas-fired power plants.

But New South Wales (NSW), Victoria, and Queensland, which are rich in coal, lag in renewables because they have been locked in to legacy coal-fired power plants. But these states are now rapidly investing in wind/solar farms and big batteries because companies are starting to retire coal power plants.

The Resilience of Batteries

It's hard to believe, but rooftop solar systems made up 29% of all renewables installed in Australia in 2023, up from 26% in 2022 (Vorrath 2024). Rooftop solar made up over 11% of Australia's electricity in 2023. In supplies to the national grid, black coal is first at 39%, brown coal and wind power are at 13%, and then rooftop solar is at 11%, ahead of natural gas at 7%.

It was also a record year for grid batteries, with investments totaling close to $5 billion in 2023 compared with $2 billion in 2022. Twenty-seven grid batteries were under construction in 2023 for a total of 5 GW/11 GWh, compared with nineteen projects in 2022 at 1.4 GW/2 GWh.

In contrast, new commitments to grid-scale electricity generation by renewables dropped to $1.5 billion in 2023 from $6.5 billion in 2022. There were only six new solar projects and no new wind projects in 2023. Squadron Energy just missed the 2023 tally but plans to deliver 14 GW by 2030, which will amount to a third of Australia's renewables goal.

Australia has had a strong history of grid batteries since the first one built in 2016 at Hornsdale, in South Australia (Palmer 2020a). It was the largest in the world and expanded in 2019 to 150 MW. Australia recognized early that a combination of solar, wind, and batteries has deep benefits. As coal-burning power plants close, federal and state governments have been putting contracts out to tender to support renewable forms of power (Parkinson 2023a).

BlackRock's Akaysha Energy is building an 850 MW/1,680 MWh Waratah Super Battery where the coal-fired Munmorah Power Station is being shut down. At the Liddell site shutdown, AGL Energy has won a tender for a 500 MW/1,000 MWh battery. Akaysha has also won a tender for a 415 MW/1,660 MWh at Orana, in the central west of NSW.

Even Denmark is getting in on the action with a 480 MWh battery, the biggest in South Australia, to be built by the energy company CIP.

The backstory in South Australia is insightful. In 2016 a hiccup occurred in South Australia's transition from coal to wind and solar, when the entire state suffered a blackout for a day or so (Palmer 2020a). Mike Cannon-

Brookes, a tech billionaire, placed a bet with Elon Musk about whether he could build a grid battery that would protect the state from future blackouts. Tesla constructed the Hornsdale battery (fig. 13-1) to store power fed by onshore wind farms that were replacing the one and only coal power plant which had been decommissioned.

Fig. 13-1. The big battery at Hornsdale Power Reserve. (Photo courtesy of David Box.)

A revolutionary idea in the search for renewable resilience, the 100 MW big battery was called the Tesla battery until it was bought by Neoen, and it was the largest in the world. It was expanded to 150 MW in 2019. One concern was how long a big battery could provide power in case of a disruption. The latest big batteries can provide full power for about four hours or half power for eight hours and can at least get most of their constituents through the night when solar power drops to nothing.

South Australia, with a population of only 1.8 million, made a remarkable comeback in renewables after 2016. It now has five big batteries and is getting 72% of its electricity, on average, from wind and solar and batteries (Carroll 2023.)

Grid Batteries Booming

Battery energy storage systems (BESS) are booming now in Australia. ("Australia's Grid-Scale Battery Storage" 2022). As of 2022, there were 800 MW of power provided by more than fifteen such grid-battery projects. More than 3,700 MW (or 3.7 GW) of new capacity began construction in 2023 (Parkinson 2023b). This actually exceeds the total capacity of wind and solar farms—by 50%. The main benefits are (1) increasing the grid capacity as coal-fired power plants shut down and (2) saving rooftop solar electricity until the evenings.

Australia leads the world in big batteries. Wood Mackenzie reported the country has announced BESS that may reach 40GW by 2032 (Kerr 2023). This is revolution-scale growth, as it amounts to almost 40% average growth year by year for eleven years until 2032. This may be compared with global growth of battery systems, due to expand from 125 GWh in 2023 to 610 GWh in 2030—an increase by 4.9 times in seven years. The corresponding Australian numbers are 3 GWh to 54 GWh, an increase by 18 times. While the difference is dramatic, what's even more stunning is that by 2030 Australia will have close to one-eleventh of the world's battery storage capacity.

Battery system costs are critical. Costs of battery modules are expected to fall by 40% in Australia by 2032. This will push down costs of overall battery systems by about 20% over the next ten years. Wood Mackenzie in their report above predicted that solar, wind, and batteries will undercut coal and gas as soon as 2028 in Australia.

The 20% lower costs of overall battery systems seems like the perfect marketing future for a transition to renewables in Australia—until the numbers are compared against China. In China, the costs of battery systems are approximately 35% lower than in Australia. Furthermore, the costs of battery systems in China are expected to drop 50% by 2032. China and Australia are both bending the arc of climate change in their battery success, but Australia doesn't make the insides of batteries. It does mine lithium, but it exports a lot of it to China, which does make batteries—hence the cost differential.

Potential Headwinds

The future does hold some concerns. One is a potential shortage of private capital to continue the expansion of wind and solar and batteries in a country limited by population and large-scale economics (Parkinson 2024). Secondly, there is a need for longer-duration batteries. Short-duration batteries, with operating times at full power up to four hours, are common in Australia. But medium durations of eight to sixteen hours are rare, and long durations of twenty-four hours or more have not been developed.

If a solar or wind "drought" occurs, it can be compensated by an adjacent wind or solar system. However, in rare cases of a double drought, the grid needs to pull on an independent system—a gas-fired power plant (non-renewable) or a pumped hydro scheme (renewable). Because these are rare cases, there is not a strong market to attract investment, and the government may have to promote and pay for pumped hydro if it wants to stay renewable.

Two criticisms of lithium-ion batteries are battery failures and fires. In March of 2024, two people died in a house fire instigated by a lithium-ion battery that went into thermal runaway. This means it overheats, gives off toxic gases, and explodes into flames. There are two causes: draining the battery too hard or overcharging it. Almost 270 battery fires were reported in NSW in 2023. The incident in 2024 was the first lethal accident, although such battery fires are relatively rare events (Nichols and Hyland 2024).

Roughly the same number of battery fires occurred in New York City in 2023, with eighteen deaths. But these were mostly batteries on e-bikes and scooters, made worse by aging of bikes purchased during the pandemic (Reyes 2024).

In regard to EVs and their low fire risk, statistics are available for Tesla cars. In 2019, the chance of fire in an internal combustion engine, per miles driven, was about nine times greater than for Tesla EVs (Lindner 2023).

Example of Quinbrook

Quinbrook, an Australian renewable energy company, has signed a huge deal with the Chinese battery company CATL (Parkinson 2023c). Their goal is to deploy 10 GWh of battery systems over the next five years using CATL's advanced storage solutions. This includes projects in Australia, the US, and the UK.

One joint project called Gemini, in Las Vegas, is one of the largest solar and battery enterprises in the US at $1.3 billion. Quinbrook and CATL's aim is to shift solar power into nighttime, as demanded by the bright lights of

Las Vegas. Commissioning plans in 2023 called for a 380 MW/1,400 MWh battery system to store the solar power.

A larger challenge for Quinbrook is the 20 GW SunCable project in Darwin, Australia, designed to transport steady green electricity continuously to Singapore via 2,700 miles of subsea cable. The battery storage time will be pushed toward sixteen hours by research on lithium-ion chemistries in CATL. The SunCable project is controlled by billionaire Mike Cannon-Brookes's Grok Ventures—the same man who made a bet with Elon Musk over building the Hornsdale big battery in South Australia in 2016.

Quinbrook is ambitious. Fifty GW of renewable and battery projects are in its huge solar and wind supply pipeline, which are enjoined with large-scale battery storage solutions that are pushing to extend the storage times.

Example of AGL

AGL is an Australian energy company with electricity, gas, solar, and renewable energy services. The AGL story is important because it's the largest coal-fired power plant company in Australia, and it says it's shutting down its last two coal power plants by 2035. AGL is a story of proactive energy transition that may resonate with utility companies in the US.

AGL has completed a 250 MW battery on Torrens Island, in South Australia. They also recently won a tender for a 500 MW/1,000 MWh battery at Liddell in NSW. Other huge batteries are in the development pipeline: a 500 MW/2,000 MWh in Tomago, NSW, and unidentified plant sites for 400 MW in NSW and 500 MW in Queensland.

But AGL is also investing in wind and solar energies: 6.3 GW of new wind and solar capacity by 2035, as well as 5.9 GW of storage capacity by 2035. The wind/solar and big batteries are hand in glove.

AGL provides interesting clues to its business perspective. First, AGL anticipates that electrification of households, especially EVs, will result in a massive shift to wind and solar.

Second, as a result, coal power stations will be the first to go, but AGL also foresees a gradual decline in demand for natural gas.

Third, and surprisingly, AGL anticipates that big batteries will make higher financial returns on its investment than solar or wind farms.

Note that Origin Energy, another large Australian electricity provider and natural gas supplier, is proceeding in lockstep with AGL. Origin's plans for 14 GW of total new capacity—wind and solar and batteries—are only a bit greater than AGL's plans.

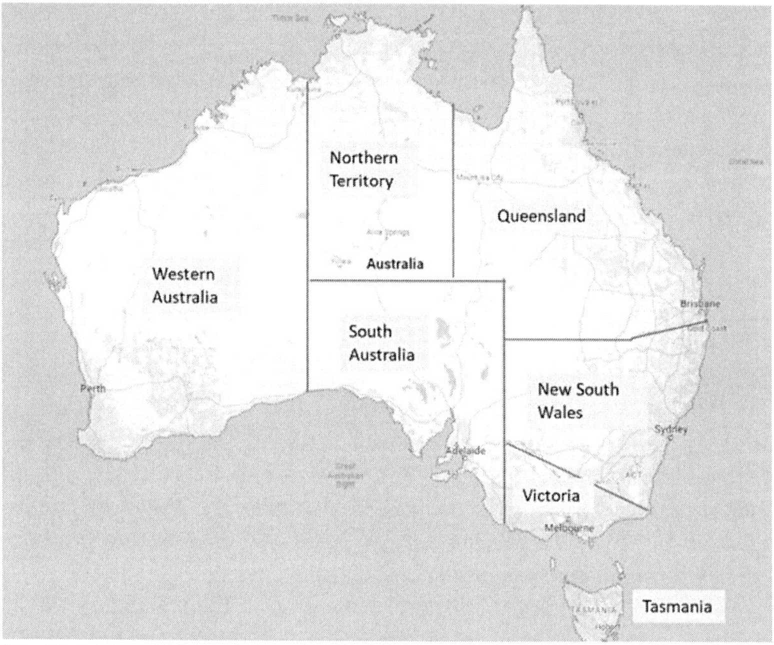

Fig. 13-2. Map of South Australia. (Google Image, SA Government).

South Australia's Template

South Australia has a unique story to tell in regard to its enviable success in renewable energies (Vorrath 2023).

When averaged over 2023, 72% of the state's electricity comes from renewable sources ("South Australia on Course" 2023). According to ElectraNet, the state could reach 100% renewables by 2026 or 2027—earlier than the self-imposed target of 2030.

Rooftop solar is unbelievable and can provide 100% of the state's demand, but just in the daytime. This meshes with the dubious distinction of the state being the driest state in the driest continent of the world.

One quote sums it up nicely: "South Australia remains at the forefront of gigawatt-scale zero emission electrical systems globally" ("Our Electricity Supply and Market" n.d.).

What to remember: South Australia foretells one possible future for the US energy transition and the potential displacement of fossil fuels by renewables.

First, South Australia boasts five wind farms, five solar PV farms, and five grid-scale batteries.

Second, 40% of homes in South Australia have rooftop solar, and thirty thousand residences have in-home batteries installed—in a state that has a population of only 1.8 million. A renewable-electrified home works like this: The home is powered from the rooftop first. Then excess electricity is fed into the home battery. When the battery is fully charged, any excess electricity goes into the grid.

Third, South Australia has shown the world what can be done ("Leading the Green Economy" n.d.). It has lifted its energy systems from 1% to 72% averaged renewable energy in just over sixteen years. On 180 separate days in 2021, renewables met 100% of South Australia's electricity demand.

Fourth, South Australia is a small state with a big vision from its government that is committed to:

> (1) large-scale renewable energy generation and storage such as wind, solar PV, solar thermal, batteries, pumped hydro, compressed air and thermal storage, (2) distributed energy resources such as rooftop solar, bioenergy and batteries, and (3) hydrogen production, use and export. (Leading the Green Economy" n.d.)

Fifth, the state is building a virtual power plant (SA VPP), which would create a network of fifty thousand rooftop solar and Tesla Powerwall home battery systems across the state ("South Australia's Virtual Power Plant" n.d.). This plan would pull together the state government, Tesla, and Energy Locals, an electricity retailer, to work to define a "community" virtual power plant.

Here Comes Tesla Electric

Homeowners, utility providers, and governments are having a tough time in Australia adapting to all the changes. An accountant is needed to figure out the financial benefits of rooftop solar plus an in-home battery. Utility providers have to ensure they continue to make a profit while an unknown number of homeowners pulls away from the grid, and governments have a hard time mandating new policies and regulations.

Tesla is planning to fill this gap and make it easy for homeowners and businesses. They want to disrupt the status quo of present electric utilities by combining rooftop solar systems, in-home batteries (Tesla's PowerWall product), and EVs into retail electric packages (Parkinson 2023e). Like similar ventures in Texas and the UK, Tesla is seeking a retailing license in several Australian states for Tesla Energy Ventures. Tesla said its goal is to power a person's car, home, and entire community with renewable electricity from Tesla (Parkinson 2023e).

If South Australia foretells the future of renewable energy, and potential displacement of fossil fuels by renewables, the oil and gas industry would be wise to look at this with eyes wide open.

Two sidebars to finish the chapter: One, Australia had 3.7 GW of new grid-scale battery capacity under construction in 2023—the hottest area of the renewables market (Parkinson 2023b). The goals are to smooth out daily fluctuations in wind and solar supply but also to shift output from the many rooftop solar systems to evening demand peaks.

Two, the biggest battery in Australia and the world is planned by Equis to be built in Melbourne (Parkinson 2023f) and is a 600 MW/1,600 MWh system, using Tesla Megapacks with lithium-ion cells, constructed from three 200 MW separates. But the most exciting extension of this system will be an additional 600 MW battery, made up of another three 200 MW separates, that uses sodium-ion cells in a flow-battery design that can supply a full load for twelve hours. This is a radical development. These sodium-ion cells are being developed with a range of different electrolytes, including vanadium and iron, and are being tested in South Korea. Equis has been given land in Australia and approvals for the total 1,200 MW of battery capacity, although a final decision on the extension will be made in late 2024.

Takeaways

Some aspects of the energy transition in Australia are truly remarkable, given the population is only twenty-six million. (For comparison, California is thirty-eight million.)

The 2017 Tesla battery in Hornsdale was a revolutionary idea in the search for resilience for renewables, and, at 100 MW, was the biggest in the world.

Australia recognized early that a combination of solar, wind, and batteries has deep benefits. Several very large grid batteries are being built on sites where coal-fired power plants are being decommissioned.

AGL is a story of proactive energy transition. They have several new grid batteries in the pipeline. AGL expects that electrification of cars and trucks will result in a massive shift to wind and solar. AGL also anticipates that big batteries will make higher financial returns on its investment than solar or wind farms.

Averaged over a year, 72% of South Australia's electricity comes from wind and solar and grid batteries. Starting from virtually no renewables, this was achieved in sixteen years. On 180 separate days in 2021, renewables met 100% of South Australia's electricity demand. These world-class achievements have occurred in a state that has a population of only 1.8 million.

A radical plan is to build a "community" virtual power plant that would create a network of fifty thousand rooftop solar and Tesla Powerwall home battery systems across South Australia. The mission of Tesla Energy Ventures is to power a householder's car, home, and entire community with renewable electricity from Tesla.

South Australia foretells one possible future for the US energy transition and the potential displacement of fossil fuels by renewables.

CHAPTER 14

CALIFORNIA LAWSUIT AGAINST BIG OIL: WHAT THE JURY NEEDS TO KNOW

Have oil and gas majors deliberately misled the public or merely expressed uncertainty about the more serious effects of climate change?

In mid-2023, five oil companies—ExxonMobil, Shell, Chevron, ConocoPhillips, and BP—along with the trade group American Petroleum Institute were named in a lawsuit brought by California Attorney General Rob Bonta. The lawsuit alleges the oil companies knew about the risks posed by climate change due to the burning of oil and gas fossil fuels but followed this up by decades of deception and lying about climate change that "forced the state to spend tens of billions of dollars to address" (Sahagún 2023).

A jury trial, imagined here, is an ideal platform to present evidence that's important to oil and gas companies as they face environmental-related damages.

Here are some other words extracted from the lawsuit allegations: "From extreme heat to drought and water shortages, the climate crisis they [oil companies] have caused is undeniable. It is time they pay to abate the harm they have caused." And "[One purpose is] to prevent the [oil] companies from making any further false or misleading statements about the contribution of fossil fuel combustion to climate change" (Sahagún 2023).

ExxonMobil is at the top of the accused list. If emissions from burning its oil and gas by parties who buy its products are included, ExxonMobil's contribution to global emissions is enormous because it is the biggest oil major in the US and one of the biggest in the world.

One goal of this book is to clarify the transition from fossil energies to renewables—how oil and gas companies are pivoting toward climate change. What evidence is likely to be brought to the table and presented in a trial about climate change? Are weather extremes worsening? What determines the urgency of net-zero by 2050? What is the responsibility of oil and gas companies, and how accurate is any deception?

Facts for a Jury to Consider

First, there is a question of the definition of *climate change*, which is generally defined over lengthy periods of time, such as twenty to fifty years. This lies in contrast to weather events, which are defined on a daily or monthly basis.

Natural events such as cycles of El Niño and La Niña occur every three to seven years, and they affect climate. In the desert of the Southwest US, generally more storminess occurs with El Niño, including intense atmospheric rivers in California in early 2024. The cause is a natural uprising of heat from ocean depths in the eastern South Pacific. However, climate is generally defined over longer periods of time.

Climate changes the earth is experiencing now are not caused by long-term cycles of the earth's orbit over tens of thousands of years—because the current rise in temperature and carbon emissions are much more rapid. Carbon emissions in the atmosphere have shot up in the last one hundred years and are now stronger than they've been in three million years. If global temperature, which is tracking with the emissions, is not controlled, it will rise by the year 2100 to a level not seen on earth for almost a million years (Palmer 2020a). This may be a twilight zone the world does not wish to be in.

Climate change is caused by man-made effects such as carbon emissions—CO_2 or CH_4. The majority of carbon emissions (about 73%) come from fossil fuels. For example, over the past forty-three years, the Arctic region has warmed nearly four times faster than the rest of the world—which has warmed by 1.2°C since the late 1800s. The warming has led to a serious loss of Arctic ice (Bamber 2022).

But wait—it's not this simple. The jury needs to know there exists a chain of events when discussing man-made climate change, as shown by fig. 14-1. This is important to understand cause and effect.

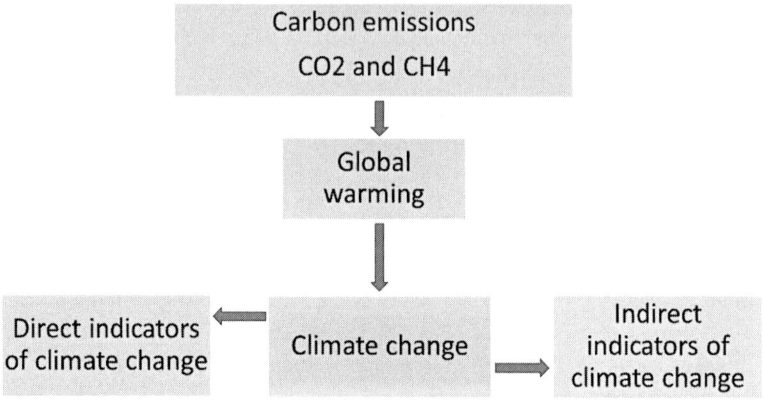

Fig. 14-1. Chain of events between carbon emissions and climate change. (Figure by author.)

Carbon Emissions and Global Warming

The top two panels of the chain shown in fig. 14-1 were studied comprehensively by Exxon in the late 1970s and 1980s and all the way into the 2000s (Palmer 2023j).

An Exxon study done in 1982 focused on global warming caused by increasing carbon emissions. Remarkably, Exxon accurately predicted that by 2019, the earth would hit a CO_2 concentration of 415 ppm and a temperature increase of 0.8°C since 1980 (see fig. 5-1).

These were serious, sophisticated studies of global warming predictions that Exxon was not afraid to publish in peer-reviewed journals between 1983 and 1984—in the *Journal of the Atmospheric Sciences* and an *American Geophysical Union* monograph (Banerjee et al. 2015).

Exxon knew the CO_2 levels were increasing over time and must have known the major contribution to this was from the burning of fossil fuels—including its own production of oil and gas. Fossil fuels today contribute 73% of all global carbon emissions.

It is clear that Exxon got *global warming* right forty years ago (fig. 5-1) and confirmed it twenty years ago (see fig. 5-2). But in later years it was alleged that Exxon in public cast doubt on the science of *climate change*.

Global Warming versus Climate Change

A jury needs to understand that Exxon's main predictions in the 1970s and 1980s were about *global warming*, not *climate change*, which is a different creature.

As an example of the confusion between global warming and climate change, a recent report said the following about Exxon:

> "New academic research lends statistical rigor to this concept by showing that the company's own climate projections, dating back decades, consistently predicted the [global] warming that was to come primarily from burning fossil fuels. . . .
>
> This core understanding—that Exxon executives knew climate change was real, but publicly cast doubt on the science anyway—was first revealed in a 2015 investigation by Inside Climate News. " (Kusnetz 2023).

But prediction of the effects of climate change from global warming is another big step (fig. 14-1) and is undercut with uncertainties (Palmer 2020a).

In years past, Exxon did acknowledge potential effects that global warming might have on *climate*, but after 1997 uncertainties received focus. Here are a few quotes by Exxon or ExxonMobil (Palmer 2023j). In 1980, Exxon executive Roger Cohen wrote:

> "There is unanimous agreement in the scientific community that a temperature increase of this magnitude would bring about significant changes in the earth's climate, including rainfall distribution and alterations in the biosphere." (Banerjee et al. 2015)

In 1997, Exxon's CEO, Lee Raymond, said:

> "Let's agree there's a lot we really don't know about how climate will change in the 21st century and beyond." (Banerjee et al. 2015)

In 2013, chief executive Rex Tillerson said:

> "The facts remain there are uncertainties around the climate . . . what the principal drivers of climate change are." (Kusnetz 2023)

A jury in the lawsuit will want to know what the evidence is for *climate change*. There are two lines of evidence (table 14-1). The first list contains *direct* indicators of climate change, while the second list includes *indirect* indicators. The evidence is much stronger for the first list than it is for the second list, despite what some of the press and climate specialists argue.

Direct indicators of climate change	**Indirect indicators of climate change**
Arctic/Antarctic ice melting	Droughts
Glaciers retreating	Wildfires
Sea level rising	Flooding
Corals bleaching	Hurricanes
Biodiversity changes	

Table 14-1. Direct and indirect indicators of climate change. (Table by author.)

Direct Indicators of Climate Change

The *direct indicators* of table 14-1 are mostly settled, because long-term data exist to confirm these trends. Arctic ice melting is well documented. But now the massive Antarctic ice sheet is also melting at an unforeseen rate, and this has potentially serious consequences for the flow of ocean currents around the globe and their potential effect on humankind ("Antarctic Ice" n.d.).

A large majority of glaciers have been retreating since 1850, the nominal start of the Industrial Revolution. The Third Pole of mountain ranges in China, Kazakhstan, Nepal, India, and Pakistan, including the Himalayas, are especially vulnerable to global warming because of their altitude. The warming of the Third Pole has been 0.3°C per decade—like the other two poles, that is three times as fast as the global average. Ten thousand glaciers have retreated, and the meltwater has formed thousands of new glacial lakes with natural dams. When triggered, these natural dams can collapse and cause outburst floods. A new study has estimated almost two hundred thousand people at risk in the Third Pole region.

A UN-sponsored report said global mass coral bleaching first occurred in 1998 and has been increasing since then (Souter et al. 2021). Roughly

14% of the world's coral has been lost since 2009. Marine scientists surveyed 719 shallow water reefs on the Great Barrier Reef in 2022. Over 90% of the reefs were bleached by warmer seawater, the sixth mass bleaching event since 1998 (Morton and Cox 2022).

Sea levels are rising in most places around the world, and the average in the recent decade is 2–3 mm per year, which amounts to 0.08–0.12 inches per year. This would mean only about eight inches total by the year 2100. But if we allow for an exponential rate of increase, rather than linear, the total sea level increase would be about thirty-six inches, or three feet (Palmer n.d.).

In one case, an optimistic projection is that by the year 2100, over four million people in the US will suffer a "disruptive inundation" in five hundred coastal areas (Schuler 2023). The report said that communities are not well prepared for disasters like this.

But, to keep this in perspective, seventy-seven years from now is plenty of time to adapt to the situation, like building six-foot-tall seawalls to protect low-lying countries where storm surges are known to damage buildings and threaten people's lives.

It would not be cheap, but a seawall would reduce the cost of regular storms and would also save lives by protecting against a disastrous five-hundred-year flood event. For example, Singapore plans to spend $74 billion over the next one hundred years to protect against rising sea levels (Lee et al. 2023). That's about $740 million every year to build seawalls.

What to remember: Climate change is not the same as global warming.

The last item of direct indicators in table 14-1 is *biodiversity changes*. The English mastermind David Attenborough has published videos and documentaries on this subject. One theme that emerges is the large number of variables that influence biodiversity changes. Climate change may be one of these variables, but it's not a dominant influencer. To say this another way, biodiversity changes are an uncertain proof of climate change in the direct list of table 14-1.

Delay in Trial

By this time, the prosecutor in the imaginary California lawsuit will have presented the *direct indicators* that look like convincing evidence for climate change. But this is not the whole story. The defense has yet to address the *indirect indicators*. They will have good arguments that the killer quad of extreme weather events, which cause many of the major problems for humanity, do not support climate change.

Let's suppose the defense will request a delay in this trial to firm up their evidence. The judge approves, and this evidence will be presented in the next chapter.

What a Jury Will Hear about Deception

Meanwhile, let's look ahead and speculate what the jury will need to know about the alleged deception by the five oil companies mentioned by the plaintiff in the lawsuit: "Decades of deception when lying about climate change."

First, there has been no oil company deception about carbon emissions causing global warming (Palmer 2023j). Exxon proved this decades ago and willingly published their findings.

Second, there were official statements made by Exxon about uncertainty regarding whether global warming caused climate change. Yes, the data show there are *direct indicators* of climate change that are worsening. But the *indirect indicators* do not show long-term evidence of the worsening of drought, wildfires, hurricanes, and flooding, even though global warming of about 1°C occurred in this period. This may have been the uncertainty of climate change that executives of Exxon were referring to.

So when the jury returns, they will have to consider the following. If there was deception, was the deception due to major oil companies honestly expressing uncertainty about the indirect indicators of climate change? Or was it caused by climate scientists, governments and media agents who ignored the long-term evidence of the killer extreme weather events?

Takeaways

In mid-2023, five oil companies—ExxonMobil, Shell, Chevron, ConocoPhillips, and BP—were named in a lawsuit brought by the California attorney general.

Climate change is caused by GHG emissions, the majority of which, about 73%, come from fossil fuels. CO_2 emissions in the atmosphere are now stronger than they've been in three million years. If global temperature, which is tracking the emissions, is not controlled, it will rise by the year 2100 to a level not seen on earth for almost a million years.

It was alleged that Exxon, in public, cast doubt on the science of *climate change*. Exxon's main predictions in the 1970s and 1980s were about *global warming*, not *climate change*, which is a different creature.

A jury will need to understand that the evidence is much stronger for direct indicators of climate change, such as polar cap ice melting and glacier retreats, than it is for indirect indicators, such as killer extreme weather events, despite what the press and some climate experts argue.

There has been no oil company deception about carbon emissions causing global warming. It was Exxon who verified this connection decades ago.

If there was deception, was the deception due to major oil companies honestly expressing uncertainty about indirect manifestations of climate change such as killer hurricanes? Or was it caused by governments and media agents who ignored the evidence that killer extreme weather events have not worsened due to global warming of 1°C in the past forty to fifty years? The jury is waiting to hear the evidence . . .

CHAPTER 15

CALIFORNIA LAWSUIT: KILLER WEATHER EXTREMES ARE NOT GETTING WORSE

Data from killer weather extremes reveal less urgency about temperatures rising another 0.3°C—and also less blame placed on big oil which produces 50% of carbon emissions.

Chapter 14 introduced an imaginary lawsuit against several big oil and gas companies. A jury in the lawsuit will want to know what the evidence is for *climate change*. There are two lines of evidence. (See table 14-1 in the previous chapter.) The first contains *direct* indicators of climate change, while the second highlights *indirect* indicators.

In this chapter, the imaginary defense will show that indirect indicators such as the killer quad of droughts, wildfires, flooding, and hurricanes have not worsened over the last forty to fifty years, on a global basis, even though the global temperature has risen 1°C. The killer weather extremes are those that can wreak havoc in the form of famines, flooding, migration, government instability, and oftentimes death.

The defense will confront the jury with this opening statement: If the four killer weather extremes haven't shown any global worsening trend over the past forty to fifty years, even though the global temperature has risen almost 1°C in this time, why should the world be worried about the next 0.3°C rise in global warming that would bring the world up to the Paris Agreement danger level?

Indirect Indicators of Climate Change

The *indirect* list of table 15-1 contains the most expensive extreme weather events that occurred in a very dangerous year. Enormous damage and massive costs to repair infrastructure and economy were dominated by hurricanes, floods, and wildfires. 2020 was a shocking year for these weather extremes.

1. Flooding, China, Jun.–Sep., $35 billion, 280 killed
2. Hurricane Laura, US, Caribbean, Aug. 21–29, $18.2 billion, 68 killed
3. Cyclone Amphan, India and Bangladesh, May 15–22, $15 billion, 133 killed
4. Severe weather (derecho), Midwest US, Aug. 8–12, $12.6 billion, four killed
5. Flooding, Kyushu, Japan, Jul. 3–15, $8.5 billion, 82 killed
6. Hurricane Eta, Central America and southeastern US, Nov. 2–13, $8.3 billion, 309 killed
7. Flooding, India, Jun.–Sep., $7.5 billion, 1,922 killed
8. Hurricane Sally, southeastern US, Sep. 14–18, $7 billion, eight killed
9. Wildfires, California, US (Complex Fire), Aug. 17–Sep. 22, $6.5 billion, 6 killed
10. Hurricane Isaias, eastern US, Aug. 2–4, $5 billion, 18 killed
11. Bushfires, southeastern Australia, Sep. 2019–Mar. 2020, $5 billion, 34 killed
12. Drought, western US, Jan.–Dec., $4.5 billion, zero killed (Masters 2021)

Table 15-1. Worldwide billion-dollar weather disasters in 2020 (Masters 2021).

For more than forty years in the US, numbers of billion-dollar disasters have been catalogued by the National Oceanic and Atmospheric Administration (NOAA) (see fig. 16-1 in the next chapter). From the numbers catalogued: The freeze count occurred only rarely. The drought count and the wildfire count were common but each limited to one billion dollars. The winter storm count, less common, was limited to one billion dollars, except for three cases where it was two billion.

The big numbers that were common through these forty years belonged to severe storms (up to thirteen events in 2020 alone) and tropical cyclones (up to seven events in 2020 alone).

The global data and US data both confirm that the killer quad of extreme weather events are the major cause of famines, flooding, migration, government instability, and oftentimes death.

These extreme weather events are what many governments, and the press, point to as indicators of climate change. But pointing is not proof—it's easy to point, but it's harder to prove. For a scientist, proof requires looking at long-term data (i.e., the statistics) to see if history shows a trend of worsening. For example, have hurricanes increased each year since 1980, or were there more Category 5 hurricanes during this period, when global temperatures increased by almost 1°C? The historical data, on a global basis, will now be reviewed briefly.

A separate examination of long-term data has been provided by Gregory Wrightstone (2017) in his book called *Inconvenient Facts*.

Global Hurricanes

Hurricanes are a critical element when looking to see if weather extremes are worsening due to climate change, because about 60% of economic disaster damages worldwide are caused by US hurricanes (Mohleji and Pielke 2014).

As fig. 15-1 attests, there is no overall trend since 1980, whether it's *major hurricanes* or *all hurricanes* (Maue 2024). Hurricanes have not been worsening in frequency or intensity.

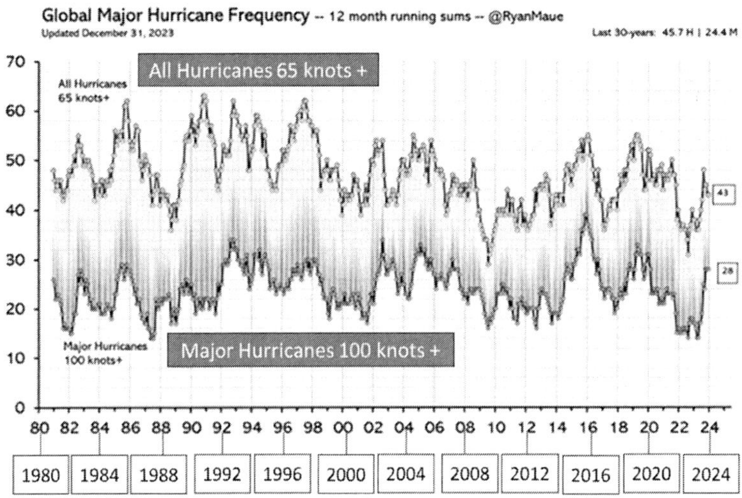

Fig. 15-1. No global trend of all hurricanes or major hurricanes in the last forty years (Maue 2024).

Global Droughts

One consequence of global warming is expected to be droughts. It's logical. But the global data of fig. 15-2 does not support this (Vicente-Serrano et al. 2022). Droughts at all three levels (mild, moderate, and severe) show no overall trend with time between 1950 and 2021, when global temperature has risen by about 1°C. In fact, there is a statistically significant decline of the percentage of land area affected by drought conditions, which is stronger with the CRU dataset but is also observed with the GPCC dataset.

What to remember: Data of the last forty to fifty years show killer weather extremes have not become worse after a 1°C rise in global temperature.

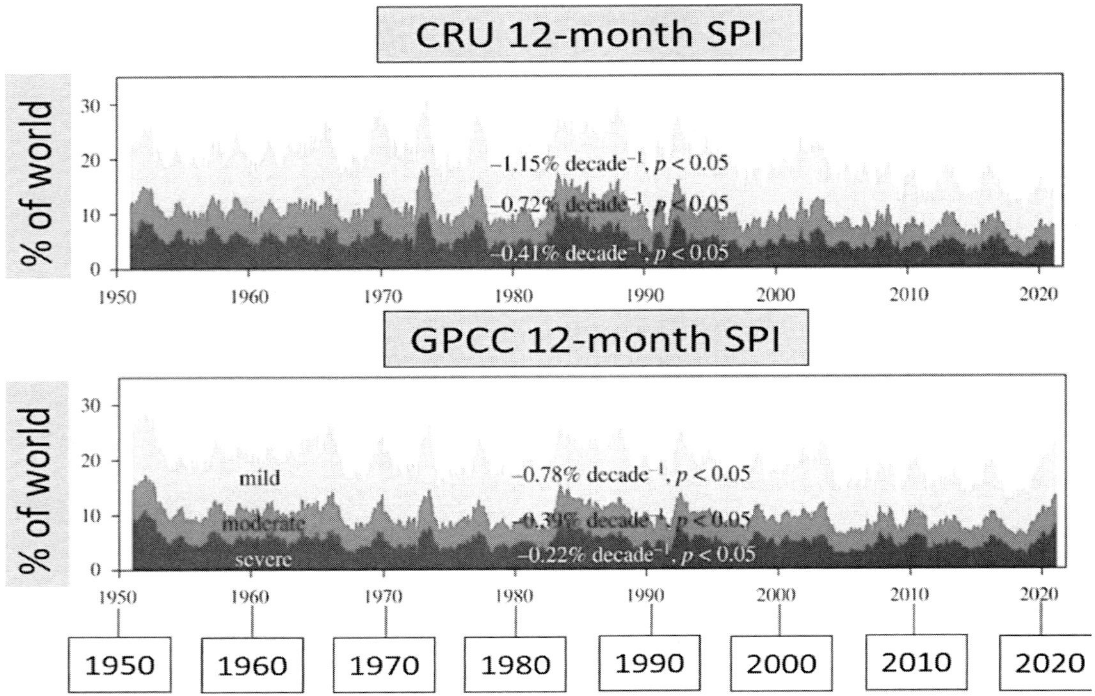

Fig. 15-2. Percentage of world surface area affected by different drought severity based on the twelve-month Standardized Precipitation Index (SPI) (Vicente-Serrano et al. 2022; licensed under CC BY.)

Global Wildfires

One way to measure wildfires is by the total area burned (Wrightstone 2017). The global data show no worsening in a recent one-hundred-year period from 1900 to 2010. In fact, across this period global area burned fell by 25%.

Another method for evaluating wildfires is by satellite measurement of carbon emissions coming from the fires ("Copernicus" 2022). This is an important and widely accepted metric. Fig. 15-3 means wildfire emissions are declining and that wildfires are not getting worse (Pielke 2023a).

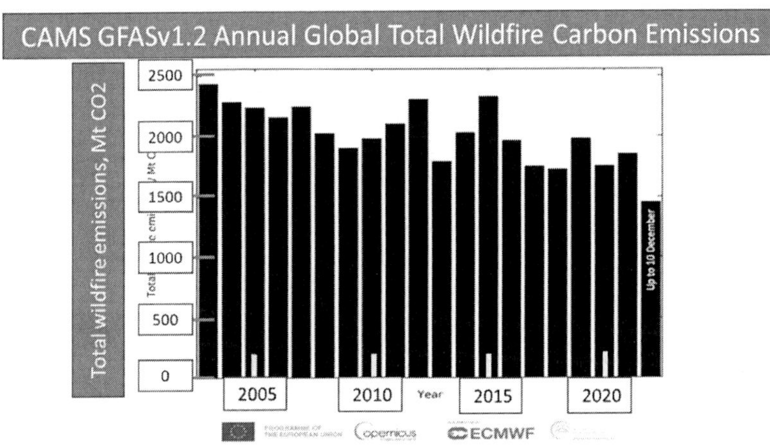

Fig. 15-3. Total carbon emissions from global wildfires. ("Copernicus" 2022; licensed under CC BY 4.0 and the ECMWF Terms of Use. Credit: Copernicus Atmosphere Monitoring Service.)

This doesn't mean that wildfires have decreased in every place. For instance, wildfires have *increased* over recent decades in the western United States, France, and Russia. But the graph does mean that wildfires have not increased *globally* in recent decades and therefore cannot be attributed to *global warming*.

Fig. 15-3 applies to global averages. On a local scale, such as in the US, wildfires do reveal a worsening over the past forty years (fig. 15-4). Therefore, there must be one or more other places across the world where wildfire statistics are not worsening but improving. This has to be so, mathematically, so that the *global average* wildfire measure of fig. 15-3 is not worsening.

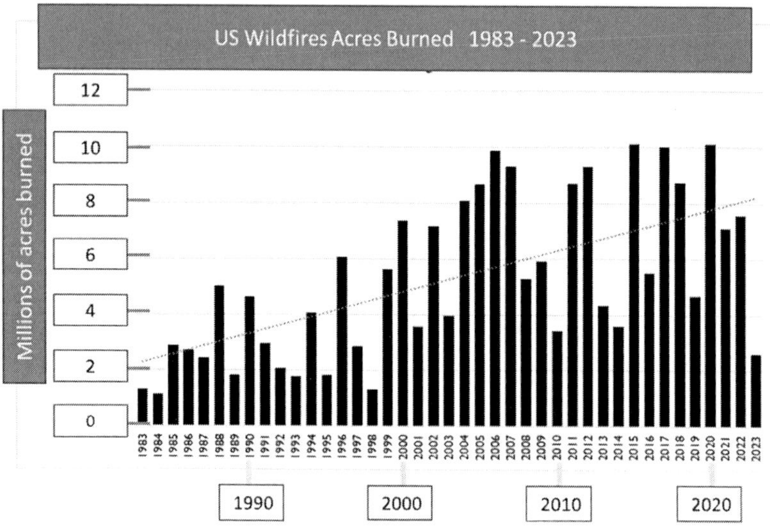

Fig. 15-4. Increasing trend of wildfires in the US since 1983. (Reproduced by permission from Roger Pielke Jr.; Pielke 2023b.)

Flooding

An insightful study from 2018 concludes the following:

> It is now well established that rising temperatures are increasing precipitation extremes. . . . Despite evidence of increasing precipitation extremes, corresponding evidence for increases in flooding remains elusive. If anything, flood magnitudes are decreasing despite widespread claims by the climate community that if precipitation extremes increase, floods must also. (Sharma et al. 2018)

The authors admit that very rare flooding cases are increasing.

But what about flood damage in the US, such as the terrible flooding that accompanied twenty to thirty inches of rain in forty-eight hours in the Baton Rouge area of Louisiana on August 11–12 in 2016 ("August 2016 Record Flooding" n.d.)? Record flooding occurred in several river basins and closed Interstate 10 to traffic for several days. Fifty thousand to seventy-five thousand structures were flooded, and thirteen people died. Has flooding, such as this once-in-a-thousand-years flood, worsened in the US?

Roger Pielke admits that *flood damage* in the US has increased since 1940, but he argues this is mostly due to greater exposure to floods by growth in the US economy that reflects a burgeoning population (Pielke 2023c). He uses gross domestic product (GDP) as a measure of this and plots flood damage as a proportion of GDP in fig. 15-5. The trend is downward, which means the US approaches to flood control have become more effective over time.

Pielke warns not to use damage costs of extreme weather disasters to argue climate change because they are biased, but to work with original climate data such as numbers of hurricanes or wildfires.

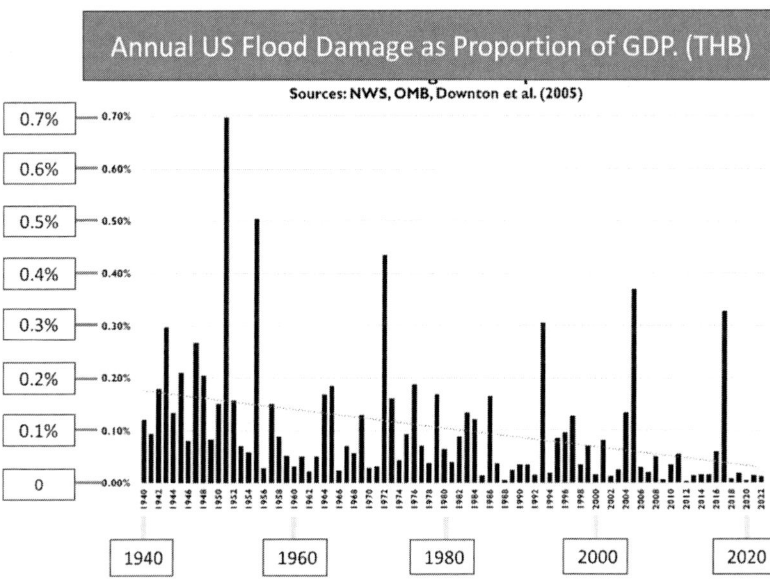

Fig. 15-5. Flooding in US since 1940, normalized by GDP. (Reproduced by permission from NOAA and climate.gov; Pielke 2023c.)

In summary, the jury will be told that none of the four killer weather extremes show any global worsening trend over the past forty to fifty years, even though global temperature has risen by 1°C in this time (Koonin 2021). The jury will be reminded from fig. 15-6 that this casts serious doubt that the next 0.3°C of temperature increase, up to the Paris danger level of 1.5°C, will bring an excessive number of catastrophic killer weather events.

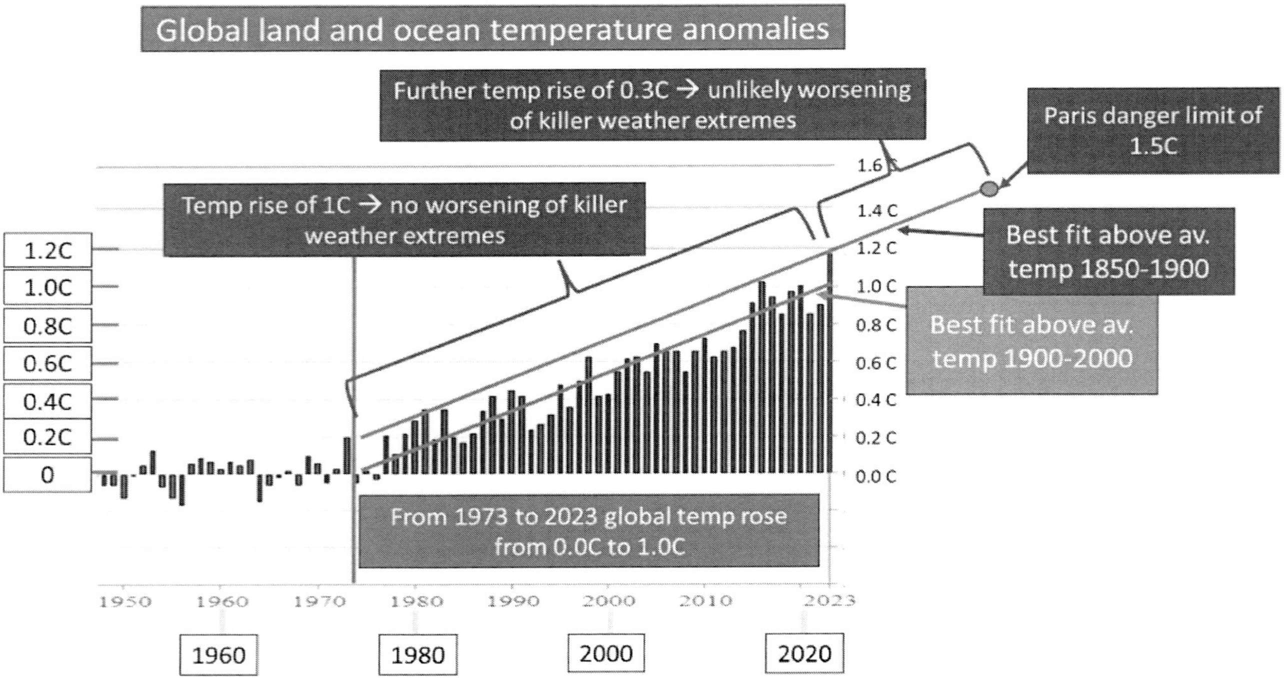

Fig. 15-6. Global temperature anomalies 1973–2023. 1°C of global warming occurred in the last fifty years. (Figure by author with data from NOAA; NOAA 2024a.).

When presented to the imaginary jury, fig. 15-6 provides an important summary and critical interpretation of the NOAA data of global temperature rise:

1) The global temperature rose by 1°C between 1973 and 2023 (lower oblique straight line, which shows temperatures above the baseline average of the period 1900–2000). It's within this period that, on a global basis, the killer weather extremes did not worsen—the global data are not sensitive to such a large temperature increase.
2) The upper oblique line is based on a different baseline average temperature: in the period 1850–1900, and before the Industrial Revolution properly kicks in. By 2023 the global temperature is up 1.2°C, only 0.3°C below the Paris danger point of 1.5°C.
3) The figure compares two adjacent periods of time and temperature. In the first period, showing a rise of 1°C, the killer weather extremes did not worsen. In the second period, showing a rise of only 0.3°C up to the Paris danger point, it's hard to argue that killer weather extremes will worsen. More likely, they will not worsen, and this lowers the urgency of enhanced climate catastrophes predicted before the Paris danger point, because the killer extremes are the major cause of famines, flooding, migration, government instability, and oftentimes death.
4) Oil and gas companies are not worsening the killer quad of weather extremes, even though the companies produce around 50% of the global GHG emissions.

Takeaways

The jury has been confronted with this conclusion: If the four killer weather extremes haven't shown any global worsening trend over the past forty to fifty years, even though the global temperature has risen 1°C in this time, why should the world be worried about the next 0.3°C rise in global warming to reach the Paris Agreement number of 1.5°C above preindustrial times?

Based on the dominant dangers of droughts, wildfires, flooding, and hurricanes, the urgency to reduce GHG emissions, though still significant, is lower than generally presented by media and the press.

The oil and gas industry is not worsening the killer quad of weather extremes, even though the companies produce around 50% of the global GHG emissions.

CHAPTER 16

WHO IS RESPONSIBLE FOR BILLION-DOLLAR EXTREME WEATHER EVENTS IN THE US?

Climate change is not responsible for increases in billion-dollar weather extremes, either in the US or in the world.

Are the killer weather extremes in the US getting worse over time, and how do they rank in 2023?

In table 16-1, Pielke has compiled extreme events that occurred in the US in 2023. This is insightful because it compares 2023 with averages from previous years, as well as the trend line of previous years. None of the four killers were above average in 2023, although two did show an increasing trend in recent decades.

A significant addition is interpretation by the Intergovernmental Panel on Climate Change (IPCC). Surprisingly, none of the killer weather extremes were attributed to climate change by the IPCC. The main takeaway from table 16-1 is that the direct indicators of extreme heat and extreme cold are attributable to climate change, but this has not carried over to the indirect indicators such as the killer quad of weather extremes.

United States Weather and Climate Extremes: 2023 Summary			
Phenomena	2023 Characterization	Time period Notes	Detection and Attribution Claimed by IPCC?
Average temperature	Average	No trend since 2000 Increase since 1895	Yes, Yes
Extreme heat	Well above average	Increasing since 2000, 1895	Yes, Yes
Extreme cold	Well below average	Decreasing since 2000, 1895	Yes, Yes
Hurricanes	Below average	No trend since 1900 Increasing since 1970	No, No
Flooding	Average	No trend	No, No
Drought	Average	No trend	No. No
Tornadoes	Average	Decreasing since 2005, 1954	No, No
Hail	Below average	Decreasing since 2005	No, No
Winds	Well above average	No trend	No. No
Wildfire	Well below average	Increasing since 1983	No. No

Table 16-1. Weather and climate extremes in 2023 and their interpretation by the IPCC. (Reproduced by permission from Roger Pielke Jr.; Pielke 2023b.)

Controversial Calls on Climate

However, there are discrepancies within the UN-based IPCC's position on disasters caused by climate change.

The IPCC has a different way of evaluating long-term global trends for weather extremes. They focus on a *climate change signal* that should appear in their data if climate change were significant (Pielke 2023a). They include many more indicators (thirty-three) of climate change than the four killer's we've discussed. But their list does include heavy precipitation and pluvial flood, extreme heat, hydrological drought, fire weather, tropical cyclone (hurricane), relative sea level, and coastal flood.

Surprisingly, only one of these, extreme heat, shows current evidence of a climate signal. Twenty-four out of thirty-three indicators have shown no evidence of climate change by the year 2023. Further, a majority (twenty, or 60%) of these indicators will not have shown a climate signal before 2050. And nineteen of them will not even have shown a climate signal before 2100. The message from this IPCC analysis is that the data do not support the urgency of climate change, as manifested by thirty-three climate indicators, including four killer weather extremes.

This is curious. Many in the media and other organizations have written boldly that the IPCC has endorsed climate change. But this powerful UN-sponsored IPCC analysis does not support imminent climate change, except for a mild statement that 40% of their climate indicators might appear by 2050.

However, some inconsistency lies within the controversy, and its source is the IPCC's *Climate Change 2022* report. Here is a direct quote:

> Widespread and severe loss and damage to human and natural systems are being driven by human-induced climate changes increasing the frequency and/or intensity and/or duration of extreme weather events, including droughts, wildfires, terrestrial and marine heatwaves, cyclones (high confidence) and flood (low confidence). (IPCC 2022)

In this statement, droughts, wildfires, heat waves, and cyclones (hurricanes) are all cited with high confidence as being driven by human-induced (e.g., fossil fuels) climate changes that increase their frequency and intensity. Only floods were cited as having a low confidence in this respect.

This inconsistency within one IPCC report is disturbing, because IPCC is an influential arm of the UN who labels climate change as an existential threat to mankind. And especially since the report is based on a large body of work and eight years of assessment by the IPCC Sixth Assessment Report (AR6), which built on information from the 2014 IPCC Fifth Assessment Report.

What to remember: The urgency of climate change has been overstated by some climate modelers, and the urgency is less than what is commonly promoted by the media.

These matters, particularly the four killer weather extremes, can have severe impacts on the human race and need to be debated. The IPCC AR6 argues that the next incremental increase in global temperature will have even more impact than the previous 1.2°C step (above the preindustrial era) because certain tipping points will be reached. These include melting of the polar ice sheets, liberation of methane from permafrost, collapse of glaciers from Antarctica into the sea, bleaching of coral, loss of biodiversity, etc. Tipping points like these might have initiated in the past two or three years, but there is no evidence for them in weather extremes that have not worsened in the past forty to fifty years.

Famous Billion-Dollar Disasters in the US

Other extreme weather events have been catalogued by NOAA in fig. 16-1. However, these are numbers of events greater than a billion dollars, and even though they're adjusted for inflation, NOAA admits they may be biased by growth in population, exposure, and economies—the global economy has more than doubled since just the 1990s. The growth of GDP is one way to look at this. The UN Sendai Framework recommends using GDP to normalize time series of economic disaster losses (Pielke 2024a). GDP growth over the years follows an upward trend that is similar to the five-year average costs (less volatile line) in fig. 16-1. This suggests that the increasing cost of major weather events over time is not an accurate indicator for increasing event activity.

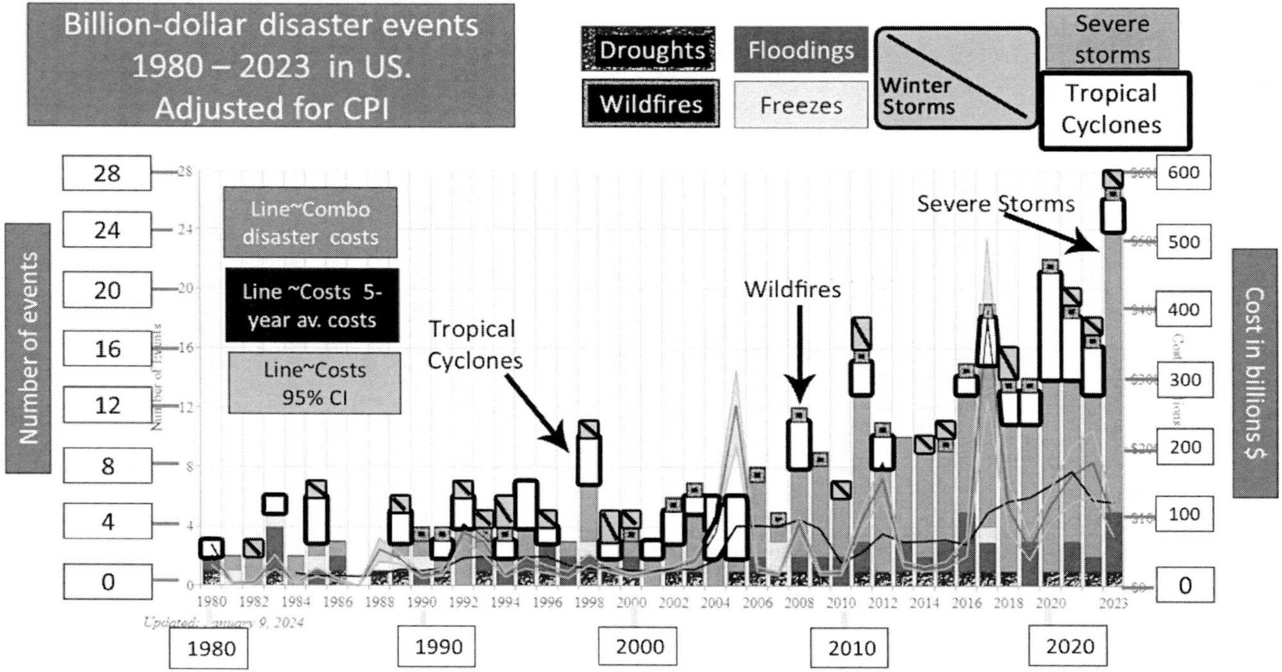

Fig. 16-1. Number of events (bars, left axis) and total cost (lines, right axis) of billion-dollar weather disasters in the US between 1980 and 2023 (adjusted for consumer price increase [CPI]) (NOAA 2024b).

Not surprisingly, Pielke (2024) has addressed this, but using global data instead of US data. Global data losses increased between 1990 and 2023, although not as strongly as US data in fig. 16-1. When the data is normalized by global GDP, the result is fig. 16-2. Costs of disasters fell from 0.25% of GDP in 1990 to 0.18% of GDP in 2023. This indicates the world has made progress in mitigating extreme weather events. Pielke points out his paper written on this subject has been downloaded twenty thousand times.

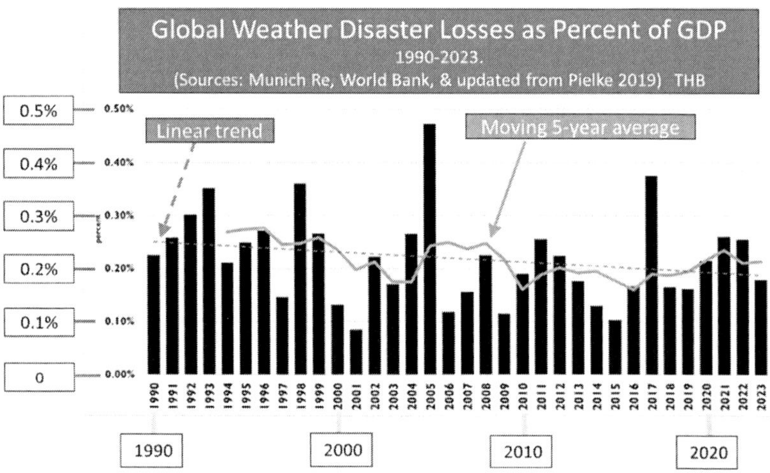

Fig. 16-2. Global disaster losses as a proportion of global GDP. (Reproduced by permission from Roger Pielke Jr.; Pielke 2024b.)

The Issue of Normalizing Weather Disaster Costs

In fig. 16-2, the disaster loss costs are normalized by GDP because GDP reflects the growth of population and infrastructure and therefore exposure to weather disasters. There are different possible normalizations in the literature, and these can lead to different conclusions.

Pielke (2024b) has discussed one case regarding statistics of hurricane damage in the US. He mentions different normalizations—in particular one by Grinsted et al. (2019) that utilizes population and wealth. The conclusion of Grinsted et al. is that in the US hurricane data, after this normalization is applied, the damage actually increases over time, which is blamed on climate change.

This is important because both the IPCC AR6 Working Group and the Sixth US National Climate Assessment (USNCA) have chosen to highlight just one normalization study: Grinsted, et al.

However, Pielke identifies a mistake that has been made in the normalization of Grinsted et al., and when corrected, the result is no effect of climate change on the cost of US hurricane damages over many years. Pielke reiterates that it's better to focus on frequency and amplitude of extreme weather events over the years, rather than damage costs, which can be controversial. The former data are consistent in their conclusion as seen in the previous chapter—no worsening of killer weather extremes due to climate change over the last forty to fifty years.

Statements from the Press and Scientific Publications

The press sometimes makes statements that seem to ignore the long-term data for the killer quad of weather extremes. One example from ABC News is as follows:

> Every tenth of a degree makes the impacts of climate change worse, Zeke Hausfather, a climate scientist at Berkeley Earth, a California-based nonprofit focused on land temperature data analysis for climate science, told ABC News. (Ebbs et al. 2023)

How is this possible, when no worsening impacts of climate change on the killer weather extremes have appeared as the global temperature rose by 1°C over the past forty to fifty years?

Another example comes from the same article:

> Just how disastrous would be the consequences the planet and its inhabitants will experience as temperatures climb? Climate researchers say the most drastic result will take the form of extreme weather events. Droughts are expected to become worse and last longer. Warmer ocean waters mean the number of strong hurricanes is expected to increase, as will the likelihood that they will quickly gain strength as they approach coastlines. Wildfires will become more intense due to hotter climates and drying landscapes. Melting ice sheets will leave some populated coastal regions underwater. (Ebbs et al. 2023)

How does this make sense when global droughts, hurricanes, and wildfires have not shown any worsening over the past forty to fifty years, when global warming rose by 1°C?

Another example of controversy surrounding this issue comes from a peer-reviewed paper published in the *European Physical Journal Plus* (Alimonti et al. 2022). The paper was about the issue of long-term data trends in extreme weather events that led to this conclusion:

> The most robust global changes in climate extremes are found in yearly values of heatwaves (number of days, maximum duration and cumulated heat), while global trends in heatwave intensity are not significant. Daily precipitation intensity and extreme precipitation frequency are stationary in the main part of the weather stations. Trend analysis of the time series of tropical cyclones show a substantial temporal invariance and the same is true for tornadoes in the USA. At the same time, the impact of warming on surface wind speed remains unclear. The analysis is then extended to some global response indicators of extreme meteorological events, namely natural disasters, floods, droughts, ecosystem productivity and yields of the four main crops (maize, rice, soybean and wheat). None of these response indicators show a clear positive trend of extreme events. *In conclusion on the basis of observational data, the climate crisis that, according to many sources, we are experiencing today, is not evident yet.* It would be nevertheless extremely important to define mitigation and adaptation strategies that take into account current trends. (Alimonti et al. 2022; italics added)

Shockingly, the paper was retracted, not because of the trends in data that showed no worsening of droughts, floods, and hurricanes, but because the conclusion said the data showed no evidence for climate change (Pielke 2023d).

This discussion has come full circle, since it started by saying there are direct indicators such as polar cap and glacier melts that prove climate change, and finished by demonstrating that indirect indicators such as the killer weather extremes haven't worsened in the past forty to fifty years, when global temperatures rose by 1°C.

Long-term data argue strongly that controlling global temperature rise by another 0.3°C (to prevent it from reaching the Paris Agreement danger number of 1.5°C) is not as urgent as the UN and the IPCC say it is. This also means that the oil and gas industry, although it causes around 50% of carbon emissions, is not to blame so much.

Following is an insightful personal comment from John Cameron, who worked in oil and gas for many years:

If the IPCC conclusions overstate the urgency of GHG reduction then this is a good thing since attainment of net-zero CO_2 by 2050 is very doubtful. But transition to carbon free energy production is still very important with respect to preserving polar ice caps, reducing acidification of the oceans, slowing sea level rise, stopping biodiversity loss, stopping the rise of global temperatures, and the impact of these on human populations and earth's ecology.

Takeaways

Inconsistencies exist within the UN-based IPCC's position on weather disasters caused by climate change. Although this is confusing, goals to mitigate climate change because of its direct effects such as polar ice melting and sea level rise should be maintained.

The now-famous graph of billion-dollar disasters in the US, increasing over the years, has an alternative interpretation that doesn't support climate change. Climate change is not responsible for increases in billion-dollar weather extremes, either in the US or in the world.

The message from the IPCC's "climate change signals" and the data of non-worsening killer weather extremes is that the urgency of climate change has been overstated by some climate scientists and the urgency is less than what is commonly promoted by the media.

CHAPTER 17

WHAT THE PRESIDENT'S PERMIT PAUSE MEANS FOR THE GOLDEN AGE OF LIQUEFIED NATURAL GAS

On one hand, LNG provides energy security and reduces emissions by displacing coal-fired power plants. On the other hand, LNG causes carbon emissions from leakage and combustion.

US was Top LNG Exporter in 2023

The US surpassed both Australia and Qatar to become the world's top exporter of liquefied natural gas in 2023, shipping an average of 11.9 Bcf/d, up by 12% compared to 2022, according to the EIA. Europe accounted for close to one half of US export volumes (EIA 2024c).

Despite this boost in sovereign wealth, the Biden administration, under its view that climate change is an existential threat, has pushed back on LNG exports (Segrist 2024). On January 26, 2024, President Biden paused all approvals to permit new LNG export projects. The US DOE says it needs to update its approval process, which is based on domestic supply, energy security, and GHG emissions. The order calls for a pause that is temporary and will be followed by a period for public response, and it will not likely be resolved until after end-of-year elections, 2024. This has caused consternation within the oil and gas industry. Here are some facts that clarify the government's actions.

One of the central issues in this approval process is methane emissions from LNG. Since LNG is methane, methane leaks play a dominant role in all facets—from its production and distribution as natural gas, to its storage, loading, and shipping as LNG. A low carbon emission footprint of LNG is critical for its worldwide acceptance as a "greener and cleaner" fuel, although it may eventually be displaced by pure renewables.

Sources of Methane Emissions in the US

There are three primary sources of man-made methane releases (fig. 17-1) (*U.S. Methane Emissions* 2021). The top source is oil and gas energy. Nearly equal to that is enteric fermentation. Third is landfills. In the US, the energy sector in fig. 17-1 is mostly methane from oil and gas operations, including (1) unburned venting of gas from oil and gas wells (about 30%), (2) leaks of gas from wellheads, storage tanks, pipelines, gas-processing plants, and refineries (about 70%). If venting of the wellhead gas is burned, the flare releases not methane but CO_2.

Enteric practices include methane from raising livestock (farting and manure) and a small amount from rice cultivation. Landfill waste includes methane from fermentation of food or liquids waste in rubbish dumps, landfills, etc.

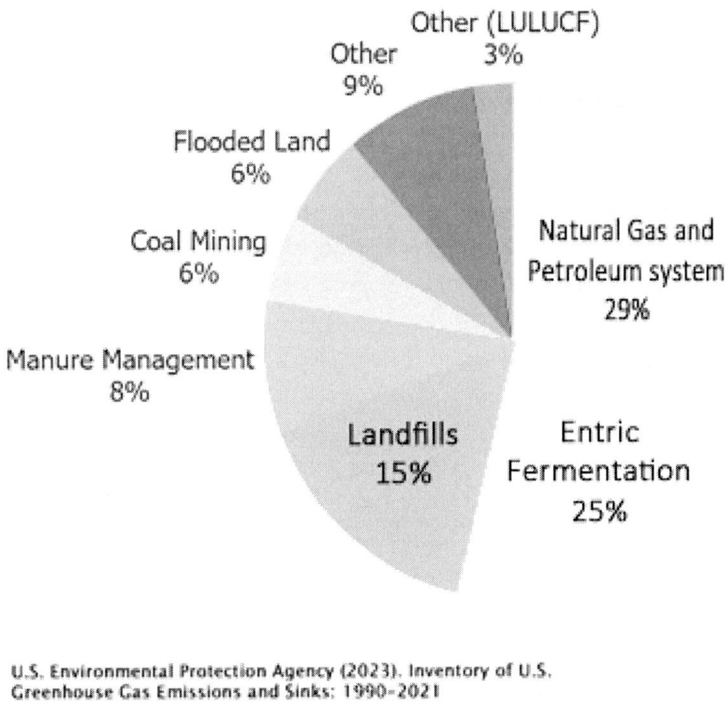

Fig. 17-1. Man-made methane emissions in 2021in the US (EPA 2024).

Leaky Footprint of Natural Gas and LNG

When oil and gas are around, methane leaks abound—from wellheads, storage tanks, pipelines, and refineries. This is important to the US government because methane heats up the atmosphere up to eighty times more than its GHG big sister, CO_2.

The UN highlighted worldwide methane emissions in a comprehensive report (Ravishankara et al. 2021). Worldwide, 60% of global methane emissions come from man-made sources. Fossil fuels contribute a third of this (20%), and oil and gas make up two-thirds of that (13% of global methane emissions). An independent estimate is 7% (Clark 2023). So oil and gas are responsible for only 7%–13% of all global methane emissions. But methane's warming effect on the atmosphere is disproportionately high, causing 25%-30% of global warming, according to the Environmental Defense Fund (Menon 2024).

What fraction of man-made emissions is the oil and gas industry responsible for? Worldwide this is one-third of man-made emissions while two-thirds of this is oil and gas, giving 22%. For comparison, in the US (as seen in fig. 17-1) oil and gas are responsible for 29% of all man-made methane emissions (*U.S. Methane Emissions* 2021). This is a higher fraction than the global-based number of 22% and must reflect the oversized role of the oil and gas industry in the US.

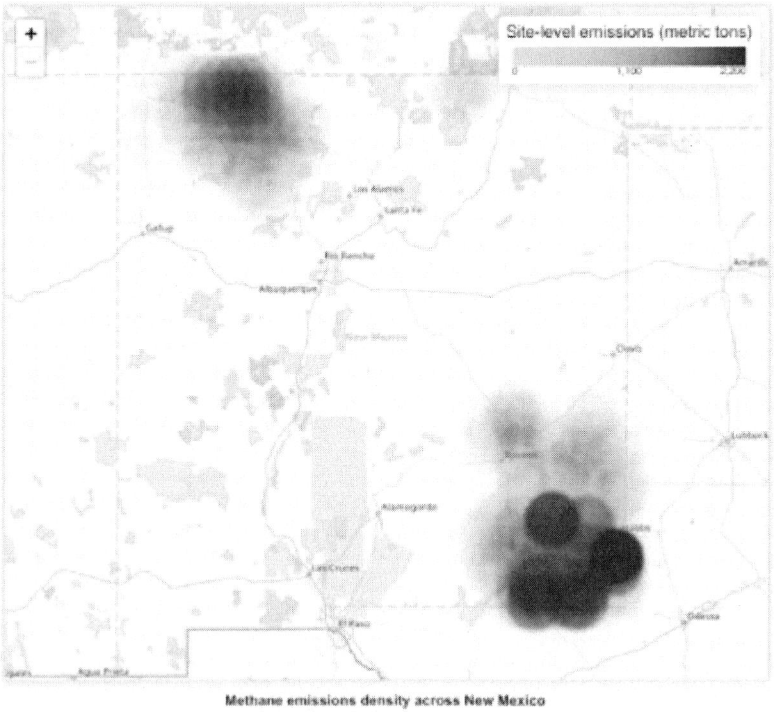

Fig. 17-2. Two hot spots of methane emissions from New Mexico in 2019 (McGee 2019).

The EDF Contribution

Nobody has contributed more to make the world aware of the methane danger than EDF (Palmer 2022k). They made systematic measurements of methane leaks at well sites by aerial surveys as well as ground-based detectors. EDF demonstrated that leakage calculations used by the EPA were deficient, meaning the EPA's numbers were too low. EDF also measured methane emissions on a statewide basis and drew attention to where emissions were rising or falling over the years.

In a game-changing publication in 2021, EDF estimated from extensive measurements that methane leaks from oil and gas operations across the US amounted to 13 million tpy (Ocko et al. 2021; MacDiarmid 2018). This was equivalent to about 1.8% of total US gas production of about 100 Bcfd. Based on a \$4/thousand cubic feet price of natural gas, this represented a loss of \$7.1 million per day, or \$2.6 billion per year, a serious amount of money.

EDF unwrapped the science. "The amount of warming over the long term is important, but so is the speed of warming," said Dr. Ilissa Ocko, senior climate scientist of EDF. "By overlooking the near-term warming from methane, we're missing an opportunity to make a real difference right now, in our lifetime. This truly is the methane moment" (Coifman 2021).

The UN picked up the baton and carried it all the way to COP27 in Scotland (Ravishankara et al. 2021). More recently, COP28 in Dubai has also formalized the methane issue. Fifty oil companies pledged to achieve net-zero methane emissions by 2030 and to end routine flaring from active oil or gas wells. These fifty oil companies amount to half of global production of oil, so their agreement is highly significant. If the pledge succeeds, it will avoid a rise of 0.1°C, equivalent to about five years' worth of temperature rise under present conditions (Gambrell et al. 2023).

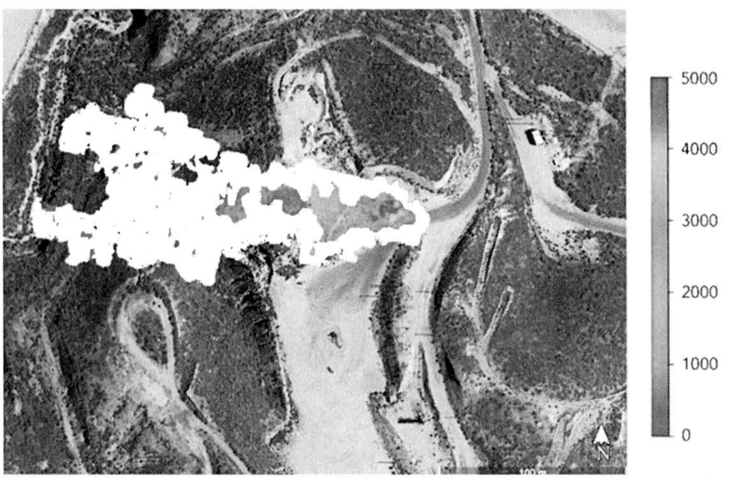

Fig. 17-3. Plume of natural gas (mostly methane) from a gas line in California. (Photo source: NASA/JPL-Caltech.)

The methane issue has also been formalized by the US EPA, which announced in Dubai its new rule for methane and CO_2 emissions from the oil and gas industry. The EPA will (1) ban routine flaring in all new wells after a two-year phase-in; (2) mandate that methane leaks from wellheads, storage tanks, processing facilities, and compressor stations be monitored properly, reported, and repaired; (3) monitor zero-emissions standards for new process controllers and pumps outside of Alaska, with one-year phase-in; and (4) identify and publish super-emitters of methane gas. Collectively, these are the strongest methane regulations in the world and will decisively lower the emission footprints of natural gas production in the US.

Another charter called the Global Methane Pledge has been signed by 150 countries to reduce emissions of methane by 30% by 2030 (Clark 2023). This charter was set up by the EU and the US in 2021.

Satellite Surveillance and Super-Emitters

EDF's emphasis on measurement of methane emissions has been invaluable to the oil and gas industry, which had previously been flying blind. Ground-based and satellite detectors have provided numbers, locations, and sources of emission plumes. The plume in fig. 17-3 enabled the operator to confirm and repair the leak.

A new study, based on measurements of methane emissions from the European Space Agency's Sentinel-5P satellite, has found that most countries have under-reported their emissions to the UN by 30% on average (Clark 2023).

The study also listed eight countries that have methane intensities of 5%–25%, much higher than the global average of 2.4%. The countries are Venezuela, Turkmenistan, Uzbekistan, Angola, Iraq, Ukraine, Nigeria, and Mexico. This cries out for better regulation to lower methane leaks.

By combining data from satellites and ground-based detectors, a French company called Kayrros has recorded, at any one time, about a hundred big methane leaks across the world as well as hundreds of smaller ones. Jean Bastin, product manager at Kayrros, explained the following:

> Over one year, those 100 leaks are releasing 20 megatonnes of methane, with around half of those attributable to the oil and gas sector and other heavy industries. This means that this sector emits an amount of methane that is equivalent to the yearly carbon dioxide emissions of both Germany and France combined. ("Mapping Methane Emissions" 2020).

What to remember: Fixing methane leaks now rather than later will lower the current rate of global temperature rise and save wasted gas and money.

MethaneSAT, a subsidiary of EDF, is a satellite that was launched by SpaceX in early 2024. It will concentrate on measuring emissions from the oil and gas industry, especially lower emissions from many sources that can add up to a serious overall level.

A correlative project called MethaneAIR, scheduled for summer 2024, will survey oil and gas basins within the US by detectors on a Learjet, although at a much smaller scale than MethaneSAT. Near real-time data, which will be free, from these two projects should enable companies to identify leaks quickly. Any leaks plugged in wellheads or storage tanks or pipelines will translate to savings of otherwise wasted gas, and benefit the bottom line. Such data will also help regulators and assist other company stakeholders, such as banking loans, in assessing how well a company is addressing climate change.

Note: both these projects will also measure methane emissions from other sources, such as agriculture and landfills.

How Fed Actions May Affect the Golden Age of LNG

The golden age of LNG has come about because of the need for energy security for the EU after Russia's war on Ukraine, as well as LNG displacing coal-fired power plants in Southeast Asia and China (Palmer 2023a). An insightful article by Bloomberg epitomizes this LNG boom (Stapczynski et al. 2024). QatarEnergy and its backers are investing $45 billion (yes, that's *billion*) to expand the country's LNG exports, although it's already one of the top three LNG exporters in the world—along with Australia and the US. A Japanese firm building the expansion has enlisted thirty thousand workers from fifty countries.

Four other large LNG expansions are planned in Louisiana and Texas: Plaquemines, Rio Grande, Port Arthur, and Golden Pass. In a few years, by adding 80 million tons of capacity, the US will separate as the top LNG exporter.

Can the Fed's LNG pause impact the LNG industry in the US? Yes. The development of an LNG project is fraught with uncertainties, especially the lengthy time frame. LNG developers in the US, such as Cheniere, try to get contract commitments from customers that last ten years or even longer—before they go about finding investors in a new project. But if they expand too fast, they risk a glut in the market. If they move too slow, there's a risk that the market might switch from LNG to coal. The Fed's pause kicks in a new uncertainty about the future of LNG.

Can the Fed's LNG pause reduce GHG emissions? The IEA has estimated emissions from all existing LNG facilities, including combustion of the LNG, and come up with 1.5 billion tpy of CO_2 equivalent, Bloomberg said (Stapczynski et al. 2024). This is 3.8% of total global emissions. For 300 million metric tons of new LNG permitted facilities, this would raise the stakes by an extra 1.2 billion metric tons of CO_2 equivalent, or 3.0% of the global total. The grand total of LNG emissions would then be almost 7% of the global total. This is what the feds have to rationalize.

A new study by Robert Howarth (2023) has looked at life-cycle emissions when natural gas is converted to LNG and then transported overseas by tanker (Olano 2024). The calculation includes production, processing, storage, and transport of natural gas before and after it becomes LNG. One baseline for comparison is natural gas that is used within the US, and this is only slightly less than coal that is used within the US.

The emissions results are normalized per unit energy source. The first results are for tankers powered by LNG, which seems logical. Compared to the baseline, the emissions are 36% higher for short trips and 55% higher for long trips. In older tankers fueled by heavy fuel oil, as is common today, the emissions are 77% higher for short trips versus 104% for long trips. These results do raise concerns about the efficacy of LNG as a bridge fuel while the energy transition shifts toward renewables.

Bill McKibben, famous for his opposition to the Keystone XL Pipeline, has protested the LNG expansion. "No one can sign a paper that says it's time to transition from fossil fuels and then permit new US projects that will add to global emissions," McKibben said (Stapczynski et al. 2024).

This is the dilemma for governments, in the US and other countries. On one hand, LNG provides energy security and reduces emissions by displacing coal-fired power plants. On the other hand, LNG causes carbon emissions, first, from methane leaks in LNG production and distribution and, second, from CO_2 when LNG is burned.

The Response to the Fed's Permit Pause

Will the Fed's permit disable the golden age for LNG? No, as projects that have been permitted won't be affected directly. Almost 50 Bcfd have already been approved for the US, which is 3.4 times greater than current capacity, so this huge boost won't be affected. This will assuage concerns from the EU and Southeast Asia about their own current plans for energy security based on cleaner LNG rather than coal.

However, the Calcasieu Pass 2 LNG project, known as CP2, has been paused, which is significant because it was due to go online in 2026 as the largest LNG export plant in the US. Surprisingly, it alone would liquefy roughly 4% of the US's natural gas that was produced in 2022. Among environmental afficionados, it's referred to as the next Keystone XL Pipeline. Concerns about the LNG industry are well laid out in an extensive article by Canary Media (Pollack 2024).

The US exports 5 Bcfd of natural gas to Mexico by pipeline. Surprisingly, some of this gas is used to make LNG in the country's own facilities. It's ironic that US natural gas will be used to support Mexico's LNG exports, which might get a boost due to the US pause in its own LNG exporting plans.

Predictably, businesses invested in LNG were critical of Biden's pause of new LNG projects. They worried about the obvious uncertainties to expanding markets in the US plus promises made to US allies such as the EU.

A Partial Solution

So is there a rational solution for the pause in permits for new LNG ventures? Yes, at least for the leaky part of LNG, which refers to emissions connected with natural gas production and distribution. These leaks are not difficult to repair, repairs can be done quickly, and the cost is inexpensive. One example is responsibly sourced gas (RSG). Another name for this is *certified gas*, which means it has been certified by an independent third party to have been produced and distributed with verified lower methane leaks.

Gas can also be certified by a company internally. In 2021, a BP vessel shipped a first LNG cargo to Southeast Asia and certified it as *carbon offset LNG*. They estimated all GHG emissions from well production to unloading terminal, and then they retired equivalent credits from their internal carbon-trading portfolio.

A natural gas certification program was announced in the spring of 2023 by MiQ ("MiQ Launches World's First" 2023). The program assesses all GHG emissions from the entire LNG supply chain: methane, CO_2, and nitrous oxides. Georges Tijbosch, CEO of MiQ, claimed the system has been used to certify almost 20% of onshore gas production in the US, and added that BP had certified all of its US onshore gas production.

Another Part of the Solution

The analysis is more complicated than this. But a recent report (St. John 2024b) from the Institute for Policy Integrity (IPI) at New York University School of Law has added two more parts of the solution. Basically, they calculated, first, how selling LNG overseas affects the domestic price of natural gas in the US. And second, how selling LNG affects the cost of climate change. It's a cost-benefit analysis that will determine if the climate cost of overseas LNG sales exceeds the economic benefits.

The first analysis was based on a 2018 study that showed selling LNG overseas would increase the price of gas in the US, but it would also boost profits to the sellers (i.e., US oil and gas companies), and this would be passed along to US consumers. The net effect would be positive.

The second analysis was in a 2022 report that calculated life-cycle emissions from gas production to shipping to unloading LNG in Asia—essentially an environmental assessment. The report found that exporting LNG would lower worldwide emissions compared with burning coal in power plants in Asia, for example.

But the new IPI report has gone further by adding other variables as follows:

> The new IPI report uses a metric known as the social cost of greenhouse gases, which encompasses wide-ranging variables including decreased human lifespans due to temperature extremes, reduced agricultural productivity due to changing precipitation, and property values damaged by sea-level rise and extreme weather events. (St. John 2024b)

After many modeling trials using different values of variables, the end result for the median case is the climate cost is close to ten times the benefit. In other words, exporting LNG is a losing proposition.

But one caveat here is that the climate costs of extreme weather events are usually high and possibly the largest of all climate costs in the modeling, as has been discussed earlier in chapter 16. Further, the killer quad of extreme weather events (droughts, wildfires, flooding, hurricanes) have not worsened over the past forty to fifty years, and so the killer extreme events cannot be attributed to global warming. Nor should their damage effects be included in the social cost used in the IPI analysis. Therefore, it seems likely the median case result will be an overestimate, and possibly by an amount that is large enough to void the conclusion of the IPI report. That is, the climate cost won't be ten times the benefit, and it could even be less than the benefit of exporting LNG from the US. The IPI analysis needs to be redone and new conclusions drawn.

These aspects of the issue will have to be rationalized against the need for worldwide energy security before the US administration decides whether to undo the pause in permitting new LNG projects.

CHAPTER 18

COP28 RESULTS AFTER TWO WEEKS OF DANCING AROUND OIL AND GAS

COP28 may go down in history as the first COP meeting to address an energy transition that includes all fossil fuels—especially a meeting that was headed by a CEO of a large oil and gas company.

The most recent UN Climate Change Conference, COP28, was held in Dubai in the Middle Eastern country of United Arab Emirates (UAE) from November 30 to December 13, 2023. About 50% of global CO_2 emissions comes from the oil and gas industry, compared with about 10% of global methane. This puts the oil and gas industry in the crosshairs of global warming.

Significant agreements came out of COP28, and what follows is a brief overview of these.

Agreement to Cut CO_2 Emissions from Future Oil and Gas Production

The really big agreement from Dubai was getting two hundred countries to agree on cutting CO_2 emissions from future oil and gas production.

A central issue was whether to *phase down* or *phase out* fossil production and consumption by 2050. The president of COP28, Sultan Al Jaber, from the UAE, had to broker the interests of the giant oil producers in the Middle East against those of the EU and the US who wanted to seriously reduce the use of fossil fuel. At the end, the agreement was to transition the world away from fossil fuels—meaning all fossil fuels—which was a first in a UN-sponsored COP meeting.

OPEC had urged its members to oppose any wording that suggests reducing oil and gas *production*, instead of oil and gas *emissions*.

The final wording issued by COP28 was "For countries to quickly shift energy systems away from fossil fuels in a just and orderly fashion. . . . Countries also are called to contribute to a global transition effort—rather than being outright compelled to make that shift on their own" (Ainger et al. 2023).

Other comments on this biggest of agreements are insightful. According to Bloomberg, Germany's climate envoy was positive about the result, saying that investors will now understand that renewable energy will be the focus of future markets, not fossil fuels (Ainger et al. 2023).

The head negotiator for Samoa, an island state, said the shift away from fossil fuels needs to have a wider basis in the whole economy, not just energy systems. Other concerns she had about the agreement were (1) that CCS might become a license for some operators to continue producing oil and gas, and (2) that natural gas, while cleaner burning, might be understood as an acceptable replacement for coal and oil for long-term usage.

Agreement to End Routine Flaring and Methane Leaks

Another agreement was that fifty oil companies pledged to end routine flaring from active oil or gas wells and achieve net-zero methane emissions by 2030.

These fifty oil companies amount to half of the global production of oil, so their agreement is highly significant. *Flaring* is the burning of excess methane at the wellhead, which releases CO_2, and this is bad. *Venting* is the release of methane from the wellhead without burning it, and this is worse because methane causes more warming than CO_2.

Other methane releases include leaks in wellheads, storage tanks, pipelines, and refineries. *How much* leakage has been controversial over the years, but the evidence was condemning after the discovery of methane clouds above basins that produced natural gas, such as the San Juan basin and Delaware basin of New Mexico in fig. 17-2.

If the pledge succeeds, it will prevent a rise of 0.1°C, equivalent to about five years' worth of temperature rise under present conditions (Gambrell et al. 2023).

But there is a wider perspective on this, as discussed in chapter 16. Even though methane's warming effect on the atmosphere is disproportionately high, the oil and gas industry are responsible for only 7%–13% of all global methane emissions.

New US Rule for Methane Emissions

The US announced in Dubai its new rule for methane emissions from the oil and gas industry.

The US EPA will (1) ban routine flaring in all new wells after a two-year phase-in; (2) mandate that methane leaks from wellheads, storage tanks, processing facilities, and compressor stations be monitored properly, reported, and repaired; (3) monitor zero-emissions standards for new process controllers and pumps outside of Alaska, with one-year phase-in; and (4) identify and publish super-emitters of methane gas. Collectively, these are the strongest methane regulations in the world.

Pledge to Triple World's Renewables

One hundred countries, out of two hundred member countries of COP28, pledged to triple the world's renewable energies installed by 2030 and to double the rate of efficiency gains by the end of the decade.

South Australia (72%) and Denmark (67%) lead the world in renewable electricity and show how it can be done. South Australia offers some remarkable statistics, as described in chapter 13.

The pledge at COP28 to triple the world's renewable energies by 2030 will be a herculean task. The most practical scenario of BP's predictions for the future of emissions (fig. 10-1) showed renewables increasing from 18% now to 26% by 2030, or 30% in the most progressive scenario (see chapter 10). These increases are not even a factor of two, let alone a factor of three.

What to remember: Two hundred countries agreed to quickly shift energy systems away from fossil fuels in a just and orderly fashion.

Still, there are solid signs of progress all over the world. Solar PV plus big-battery storage is the best renewables combination in the market right now. For example, the biggest US solar project came online in January 2024 in the Mojave Desert of Kern County, California (Wesoff 2024). The megaproject, called the Edwards & Sanborn project, contains two million solar panels and 120,000 batteries, to store electricity until the sun goes down. The capacity of this project is 875 MW and 3,287 MWh, which implies the battery could run at almost full capacity for four hours if needed in the event of a power outage. Electrical power is provided to a couple of large utilities plus the city of San Jose, plus Starbucks, which just proves again that Starbucks is everywhere.

One other minor agreement at COP28 relates to roughly 30% of all methane emissions that comes from livestock, mostly from burps and farts and manure. Six of the world's biggest dairy companies will start reporting their methane emissions by mid-2024 and will write action plans to clean up their emissions by the end of 2024 (Douglas 2023).

Meeting all the COP28 goals will require tools to measure and charts to plot changes in energy systems that will be practical, reliable, and transparent to stakeholders such as oil and gas companies, bankers, investment houses, regulators, and governments. This will be challenging (McDonald 2023).

CHAPTER 19

CAN THE WORLD REACH NET-ZERO BY 2050 WITHOUT RADICAL NEW TECHNOLOGIES?

Oil and gas companies are starting to pivot toward climate change, but how far should they pivot? Try to diversify away from oil and gas production, or stay with business as usual and hope for a moon-shot solution.

Oil and gas companies are starting to pivot toward climate change. But how far should they pivot? That is what this chapter is about. Should companies try to diversify away from oil and gas production to limit the carbon emissions they are responsible for? Or should they gamble on business to continue as usual, hoping for a creative or even a moon-shot solution that emerges from the immense creativity bubbling all over the world?

The story in this chapter begins with an eminent company that prides itself on analyzing data and forging conclusions—Rystad Energy. Once a year they present results and conclusions from a broad and deep investigation of data related to the energy transition. Their bottom line is that net-zero emissions by 2050 is an achievable goal.

The story continues with an independent analyst who is not directly connected with oil and gas business but has been writing about fossil energy and climate change for over twenty years. Roger Pielke calls himself the Honest Broker. His conclusion is that reaching net-zero by 2050 is not plausible the way the world is going.

There are many other voices whose conclusions lie in between these two bookends, as one would expect. One such is the IEA, which has done a vast amount of study and offers strong advice to the oil and gas industry to reduce, as soon as possible, its oil and gas production and carbon emissions.

The oil major BP has laid out thoughtful scenarios for future oil and gas consumption as well as renewable energies and hydrogen, based on prodigious amounts of data and analysis. Their viewpoint is valuable.

Rystad Energy—The First Bookend

The final session of Rystad's Energy Transition Day for 2023 was led by Rystad's CEO, Jarand Rystad. He stated he was optimistic that net-zero for the 1.6°C pathway could be achieved by 2050 (fig. 19-1). This figure can also be seen enlarged in fig. 11-1.

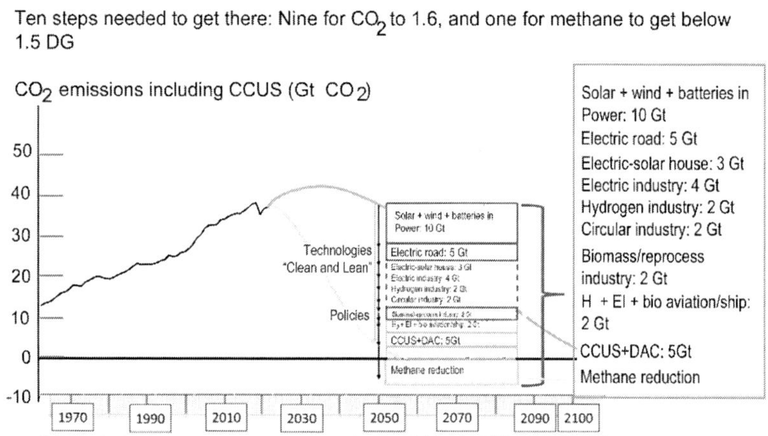

Fig. 19-1. The left projected line is for the 1.6°C climate pathway. The right projected line is for the 2.5°C climate pathway. (Rystad Energy's Energy Transition Marathon, May 2023.)

The rectangular bands represent decarbonization sectors. The large band at the top is for decarbonization due to solar, wind, and batteries and is estimated to be 10 Gt out of a total about 40 Gt to be decarbonized. This first band is a huge 25% of the total. The second band is 2 Gt due to EVs (cars, trucks, etc.). Rystad thinks these bands are definitely achievable. In fact, the CEO opined that in fifteen years we will have forgotten how to drive gasoline vehicles.

The dashed band means a little more uncertainty in the following contributions: rooftop solar, 3 Gt; industry electricity, 4 Gt; hydrogen in industry, 2 GT; circular industry, meaning waste to energy, 2 Gt.

Then comes biomass and the reprocessing industry at 2 GT, which is more certain.

The bands at the bottom of fig. 19-1 have perhaps 50% probability of success, according to Rystad. These include fuels for aviation and shipping, which may acquire a mix of hydrogen, biomass, and electricity to save up 2 Gt of emissions.

These bands also contain CCS and DAC, which together amount to 5 GT, but there's a problem here.

Fig. 6-3 in chapter 6 shows that sequestering 4 Gt/yr of emissions by CCS would result in 2°C global warming by 2050. It would take 8 Gt/yr to get on the 1.5°C pathway. According to Rystad, these scenarios would require a global capacity increase by one hundred to two hundred times the present capacity of 40 million tpy (Abramov and Lam 2023). This would represent something like a 20% growth in CCS year over year until 2050—a scenario that's hard to fathom (Palmer 2021b). The carbon capture industry would have to grow as big as the present oil and gas industry by 2050, according to Rystad. This much growth, and its costs, makes this scenario unrealistic.

There is also a problem with hydrogen, as discussed in chapter 11. There it was concluded that by 2050 the principal use of hydrogen will be for hard-to-decarbonize sectors that amount to less than 15% of global energy consumption. Hydrogen fuel is garnering lots of press and PR and excitement about a potential new industry, but applications are limited, it's expensive, and profits are uncertain.

It appears that Rystad's optimism needs to be tempered by impractical factors that involve CCS and hydrogen.

The last piece of the puzzle is about 10 Gt that would come via methane reduction from natural gas leaks in wellheads, pipelines, storage tanks, and refinery facilities. This band should have a high level of success because in the past three years, methane has been fully recognized as a very damaging GHG as new methods of observation, including remote sensing and satellite detection, have come to the fore. There is optimism, too, because recently China put in place a plan for methane reduction, and the EU agreed on methane-reduction laws.

Another observation from fig. 19-1 is if the bands of numbers are not achieved, this would mean a 2050 temperature rising to something more like 2.0°C.

An update by the CEO of Rystad in late 2023 is insightful ("Note from the CEO" 2023). The only way to reach net-zero by 2050 is, first, by reducing methane emissions by 30% by 2030 (the sprint) and, second, by limiting CO_2 emissions to 650 Gt by 2055 (the marathon race). This implies cumulative renewables would have to grow from 4 Terrawatts (TW) today to 11 TW by 2030 (an increase by almost three times). Although wind has fallen behind, solar has zoomed up, particularly in China, and Rystad believes the goal is still achievable.

Roger Pielke—The Second Bookend

This analysis is based on global energy data collected by the Energy Institute *Statistical Review of World Energy* (formerly the BP *Statistical Review*). An independent analysis of the data has been done by Roger Pielke (2023e).

Fig. 19-2 shows that global renewable energy increased every year from 2010 to 2022. Fossil energy increased every year, except in 2020, when the pandemic hit. But this was compensated by an equivalent increase in 2021. While this data testifies to a growing demand for energy globally, it doesn't indicate a transition from fossil energy to renewables has taken place.

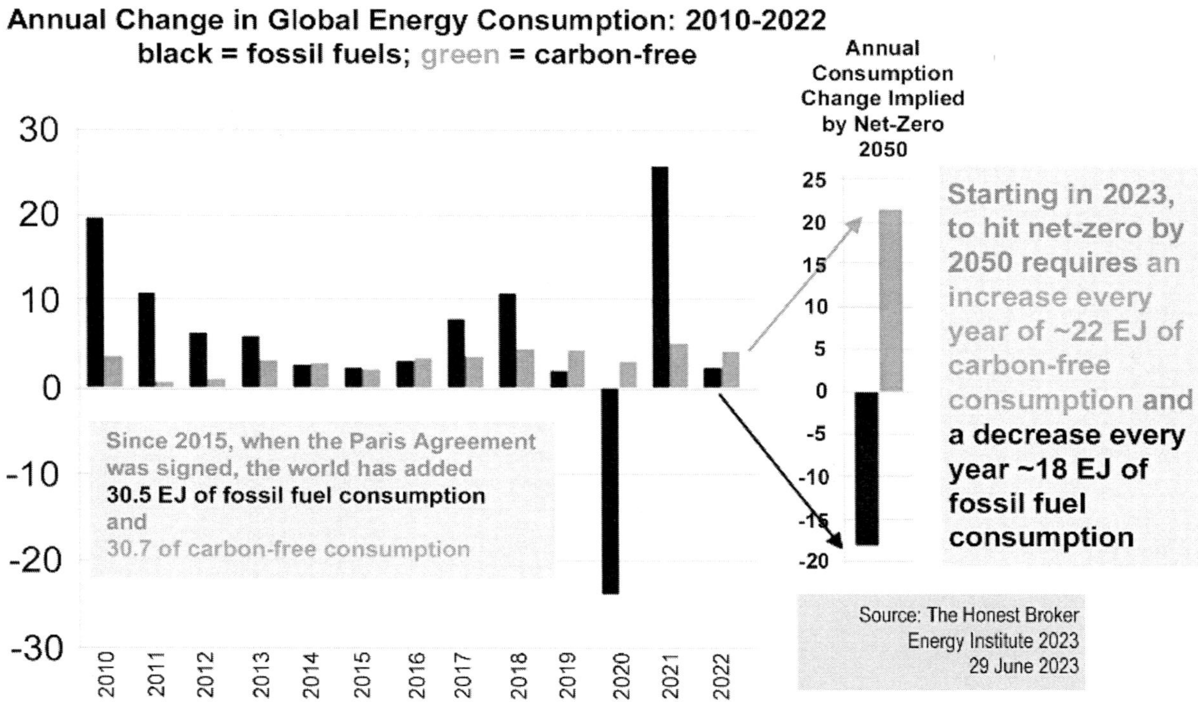

Fig. 19-2. Annual changes in global energy consumption from 2010 to 2022. Black is fossil fuels. Gray is carbon-free renewable fuels. (Reproduced by permission from Roger Pielke Jr.; Pielke 2023e.)

To get to net-zero by 2050 would require two things: first, fossil energy consumption must stop (to stop the emissions), and second, all this fossil energy must be replaced by renewable energies to meet the overall energy demand. A subtle key is that carbon emissions won't stop unless fossil energy consumption stops, no matter how much renewable energies increase.

Pielke offers a startling equation: Achieving net-zero fossil fuels by 2050 would require the construction of one nuclear power plant per day (average 1.75 GW capacity)—starting today and continuing every day until 2050. An alternative would be the installation every single day of two thousand wind turbines. (Three MW is an average turbine's capacity.) Pielke concludes it would be a gargantuan task to get off fossil energies by 2050.

To reach net-zero by 2050 would require renewable energies to increase by about 22 exajoules (EJ) every year through 2050 (fig. 19-2). And simultaneously, fossil energies would have to decrease by about 18 EJ every year. The difference between 22 EJ and 18 EJ would account for growth in global energy demand.

One other scenario would deploy CCS to remove and bury emissions from a fraction of the 18 EJ/year of fossil fuels that need to be reduced. This would allow some fossil energy consumption to continue.

But there is a caveat. Fig. 19-1 says that CCS may be needed to save 5 GT maximum out of a total 40 GT that needs to be decarbonized. This is 12.5% of the total. Rystad also argues that such an endeavor by CCS would mean a grandiose new CCS business that would have to grow by 20% year over year until 2050, when the size of the business would exceed the present oil and gas industry—an impractical picture.

Recall that the oil and gas industry produces about 50% of global carbon emissions. For CCS to expand its removal fraction from 12.5% to a larger number like 50% is completely implausible. CCS might be able to save a small fraction of fossil fuels (possibly 5%) but not much more.

International Energy Agency (IEA)—An Anxious Voice

The IEA was established in 1974 as an intergovernmental organization that collects data on global energy, analyzes it, and provides policy recommendations. Membership includes thirty-one countries and thirteen association countries that together consume three-quarters of all global energy.

A serious and confrontational report came out in November 2023 (meaning confrontational to the oil and gas industry). Here are parts of the executive summary (IEA 2023):

- The oil and gas industry provides roughly 57% of global energy but about 50% of GHG emissions. The industry employs twelve million workers. But the clean energy investments within the industry total a feeble 1%. (More than 60% of this comes from four companies out of thousands.) The IEA said the industry is a marginal force in transitioning to net-zero emissions.The IEA set some hard goals for the oil and gas industry. One is that they have to cut emissions from their operations by 60% by 2030. (Rystad said 30%.) Another is that they need to put more of their CapEx toward renewable energies. Now it's a miniscule 2.5%, but it should be pushed up to 50% by 2030. Third, oil and gas should throw out the concept that CCS can remove all the emissions from continuing oil and gas production. CCS won't save the oil and gas industry, as stated earlier in this chapter. The IEA argues that over 30 Gt/yr of emissions would have to be buried by 2050, but that is ridiculous. The cost would be over $3.5 trillion a year until 2050. This is equal to the entire oil and gas industry's annual revenue in recent years.

One touch of optimism: if the world does reach net-zero by 2050, meaning the energy supply is carbon-free, then maybe a third of this energy would come from sectors that the oil and gas industry has solid experience in. The skills and resources that would carry over are in the areas of offshore wind, liquid biofuels, biomethane, hydrogen fuel, and geothermal energy.

One of the more sobering scenarios from the IEA: the return on capital. This was 6%-9% for oil and gas over 2010–2022, while it was 6% for renewable projects—and lower profits are an obvious reason why oil and gas companies are reluctant to jump ship to more renewables. Expected energy transitions toward net-zero will hurt the bottom line of companies that are focused on oil and gas. The danger is that oil and gas production becomes a riskier business (think stranded assets) as the world gets closer to 2050 and net-zero emissions. A company will want to avoid production that has a long lead time, because a steep energy transition may trip up the completion of long-term goals. An exponential rise in EV cars and trucks and buses could bring on such a steep transition.

Of course, production of oil and gas is needed for energy security during the transition, and presently the industry invests US$800 billion per year. But the IEA says this would have to be half as much—only $400 billion per year—by 2030 to reach the 1.5°C scenario by 2050.

What to remember: Carbon emissions are reduced by scaling down fossil fuel consumption, not deploying more renewable energies.

But there is some good news from the IEA. A new report called *Renewables 2023* says the world has a good chance of tripling renewable energy capacity by 2030, a goal that was set at COP28 (IEA 2024). This capacity grew by 50% to 510 GW in 2023, with solar PV bringing three-fourths of this. China was dominant, as they installed more solar PV in 2023 than the entire world did in 2022.

In the report, renewable energies under existing policies are predicted to grow to 7,300 GW by 2028. Ninety-five percent of this increase is solar and wind. By 2025, renewables will overtake coal as the main source of electricity. But it will not be enough—renewable growth up to 2030 will only be 2.5 times 2023's capacity, not three times, the goal agreed upon at COP28.

BP—A Deep Dive into Data-Based Analysis

BP has presented its highly regarded *Energy Outlook 2023* (2023). The new perspective emphasizes the Russian war in Ukraine. Fig. 10-1 of chapter 10 shows BP's scenarios for the future, in which the Russian war in Ukraine is likely to swing the pendulum away from "business as usual" and toward the Paris goal of "net-zero emissions" by 2050 (BP Energy Outlook 2023). This scenario, called *Accelerated* in fig. 10-1, can be taken as BP's best guess for the future.

In contrast to the current enthusiasm in the US oil and gas industry, BP forecasts a *drastic* reduction in fossil energies from 83% of total energy supply now to 28% by 2050. Coal will fall the most, oil will fall, and natural gas may or may not fall. BP says natural gas is an X factor because of an ever-rising demand for LNG around the world.

BP projected that renewable energies, such as wind and solar and batteries, will rise from 18% of total energy supply now to 58% by 2050.

Side by side at 2050, fossil fuels will be at 28% and renewables at 58%—close to double the fossil fuels.

Technology Needs

Bill Gates, in his 2021 book, *How to Avoid a Climate Disaster*, emphasizes the need for innovation. He adds several ideas for government to do, particularly via R&D funding, which they are good at:

1. Quintuple clean energy and climate-related R&D over the next decade.
2. Make bigger bets on high-risk, high-reward R&D projects.
3. Match R&D with our greatest needs.
4. Work with industry from the beginning.

Gates has his own foundation that provides funds for R&D, so he has considerable experience in funding research.

His book was published in 2021, the same year the US passed the infrastructure bill, while the IRA bill was passed into law in 2022. Both bills committed massive amounts of government funding in the form of climate-related grants and investment tax credits that added up to roughly $479 billion (see chapter 6).

Green hydrogen.	Air Products; Plug Power; Neom (Saudi Arabia).	7,000 tpd from 100 plants; 2,000 tpd for fuel cells; 600 tpd from 2026.
Grid-scale electricity storage.	Waratah Super-battery, by Akaysha Energy, Australia.	850 MW1680 MWh (biggest ever).
Carbon capture.	Occidental's Stratos.	Will be 0.5 million tpy then 1 million tpy. Maybe 100 plants in US, 1,000 in world.
Advanced biofuels.	Virgin Atlantic: first Boeing 787 flight across Atlantic using SAV.	Only 0.1% of supply today so cost too high (3-5 x).
Zero-carbon cement: cement is 6-7% of global carbon emissions.	Best route for cement decarbonization is uncertain.	In US, Sublime Systems scaling to commercial.
Zero-carbon steel.	H2 Green Steel, Sweden, 2025 rollout.	Hydro for clean power + electrolysis.
Zero-carbon ammonia/fertilizer.	CWP Global; BP in Australia.	20 million tpy; 10 million tpy.
Next-generation nuclear fission. Small Modular Reactors (SMRs).	In China, Huaneng first working in world.	SMRs in Utah, US; Funded by DOE, Bill Gates.
Geothermal energy.	FORGE in Utah; Fervo in Nevada.	Uses staged fracking from shale revolution.

Table 19-1. Technologies needed to reduce GHG (adapted). (Table by author; Gates, 2021.)

Table 19-1 has been adapted from Gates' book with examples and companies and results added in each area. Clearly, there has been substantial progress in many of these technologies, and progress will expand as funding from the two new US laws is distributed. Bill Gates would be happy.

Innovations for Making Carbon-Free Electricity

Following are some more notes on the future of a few technologies, most of which are an extension of table 19-1.

Nuclear Fission

This is the only technology that is carbon-free and reliable in its output if used at any time anywhere in the world. And it can be used on a large scale. France's power is 70% nuclear fission. Nuclear power kills far fewer people than car crashes do.

For twenty years the US DOE has funded the concept and development of a small modular reactor (SMR) called the NuScale Power Module. The advantages are that it's safe, cheap, scalable, and carbon-free. It's the only SMR to receive design approval from the Nuclear Regulatory Commission. Less than one hundred feet tall, the module is a fifteen-foot-wide cylinder that sits in a bath of water below ground level. It can produce 77 MW of electricity, which can power sixty thousand homes. The goal is to have it up and running in Idaho by 2029.

But a glitch to this picture has emerged in Utah, where six 77 MW NuScale reactors were planned to be built (Parkinson 2023g). The customers, mostly municipal utility companies, have refused to pay a recent cost jump of over 50% to $89/MWh, in spite of a healthy subsidy of $30/MWh from the federal government.

A second glitch from Utah concerns Rocky Mountain Power, a subsidiary of PacifiCorp. PacifiCorp had partnered with TerraPower, a company founded by Bill Gates, for a 345 MW demonstration called Natrium in Wyoming which could break ground in 2024. In a Natrium plant the fission reaction stores heat in liquid sodium metal rather than water. (Fitzpatrick 2024). Natrium is a smaller and cheaper design that works a lot faster than a conventional reactor (1,000 MW), but is larger than a typical SMR. But PacifiCorp has suffered some big insurance losses, and backed off its plan to build two other Natrium plants in Utah that would have been the first nuclear enterprises in the state. As a result, Rocky Mountain Power has canceled previous plans to shut down two coal-fired power plants in 2036 and 2042, and switch to renewable nuclear energy.

A study by the Commonwealth Scientific and Industrial Research Organisation (CSIRO) has concluded that "The LCOE [levelized cost of electricity] cost range for variable renewables with integration costs is the lowest of all new-build power technologies in 2023 and 2030" (Graham et al. 2023). The costs include solar or wind production, transmission, and storage of electricity. After 2030, CSIRO finds SMR nuclear energy would be at least three times more expensive than renewable energies in Australia.

Despite the negatives, an SMR startup called Oklo is contacting Permian operators and has a letter of intent with Diamondback Energy, a large independent. The electrical supply can be dubious in west Texas, and while oil and gas companies want to lower their emissions, switching from diesel engines to nuclear may be more reliable than switching to electricity. The deal is years away from commercial application, but these ultra-small SMRs that will produce maybe 15 MW are of some interest (Wade and Ferman. 2024).

For nuclear energy in whatever form, success is always uneasy when terrible accidents occur, such as Chernobyl in Russia in 1986 and Fukushima in Japan in 2011. And there is an ever-present worry about nuclear waste disposal in the US (Palmer 2023m).

Nuclear Fusion

One scientific application of nuclear fusion has been at Lawrence Livermore Laboratory, in the San Francisco Bay area of California. There, 192 lasers were used to confine, compress, and heat plasma by blasting a $1 million pellet of mixed hydrogen isotopes (Purtill 2022). The results were always the same—until an announcement at the end of 2022 (Palmer 2022l). For the first time, *energy out* (3.1 MJ) was more than *energy in* (2.1 MJ). It was a genuine breakthrough. The temperature achieved was 3 million degrees Celsius, a sunlike temperature.

Mankind has finally replicated the sun's source of light and heat. At about 15 million degrees Celsius, the gaseous interior of the sun is compressed under tremendous pressure—a teaspoon weighs 750 grams or 1.65 pounds. To replicate the sun's interior conditions in a laboratory on earth and to achieve breakeven (energy out more than energy in) is an impressive feat.

However, to become commercial is a different proposition. This would require many, many fusion events per minute and would require a laser that is thousands of times more powerful. Plus the cost would have to be a million times cheaper. In a word, this one success, though inspiring, is not remotely close to practical application.

Nuclear fusion creates three to four times more energy than nuclear fission. That is one part of the dream. Another part of the dream is that there are no nuclear waste products to dispose of—fission waste products can take hundreds or thousands of years to decay. But it's only a dream. Carbon-free nuclear fusion won't put the oil and gas industry out of business by 2050 and maybe not even by 2100.

Yet there are start-ups that challenge the conventional approach discussed above. One is called Helion, which uses plasma accelerators to form two plasma clouds that are forced by magnets to collide. The heat and pressure create a pulsed fusion reaction that directly produces electricity as well as heat. Fusion power comes from a string of pulsed nuclear fusion reactions, rather than one big self-sustaining reaction. This would lead to a cheaper and smaller fusion reactor that could fit inside a railroad container.

Helion says it will start production in 2024 and even has a sales contract with Microsoft to be filled by 2028.

Geothermal Energy

Geothermal energy did supply the US with 3,700 MW in 2022, but this is only 0.4% of the total supply. The DOE said it could supply thirty times this amount, or 90,000 MW (90 GW), by 2050.

Two recent projects are testing and proving that shale technology can be adapted to extract energy from hot rocks deep underground. The story begins with the shale revolution in 2003, when the oil and gas industry discovered that a horizontal well two miles long and fracked up to forty times along its length can make commercial flows of oil and gas in ultra-tight shale (Palmer 2020a). In fact, the shale revolution reignited the US oil and gas industry so that the US became independent in oil and gas production—for the first time in sixty years.

Project FORGE

The US DOE has funded a project called FORGE, in which hot granite rock is being drilled and fracked using the best oil and gas technology (Palmer 2022m). An overall goal is to test commercial application.

In a commercial setting, cold water would be pumped down one well and pass through an array of hydraulically created fractures, acquiring heat in the process. The heated water would be produced to surface through a production well (fig. 19-3). At the surface, standard geothermal technology would drive turbines that generate electricity. The produced water, after heat has been taken out, would be recirculated.

John McLennan of the Department of Chemical Engineering, University of Utah, is the co-principal investigator for this DOE project.

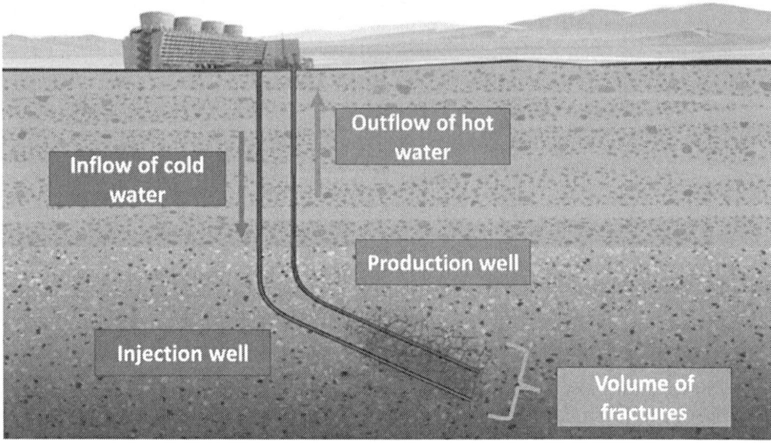

Fig. 19-3. Injection well drilled on the left and production well on the right are connected by fractures in granite. (Reproduced by permission from Utah FORGE; Forge Utah 2021.)

The production well was drilled in mid-2023, parallel to the injection well but offset three hundred feet vertically above that well (fig. 19-3). It was drilled to intersect microseismic clouds from three medium-size hydraulic fractures done in the injection well in April 2022.

Fracturing stages are planned to be pumped with varying clusters per stage—up to eight clusters which is commercial scale. Proppant will be pumped in all stages (100 mesh and 40/70 mesh), and the treating fluid will be slickwater or crosslinked guar.

The FORGE site will not produce power. It is intended to be used to test and develop technologies that will promote the commercialization of this type of geothermal energy.

Project Red

In the fall of 2023, Google and geothermal developer Fervo Energy said its Project Red is supplying 3.5 MW of electricity to NV Energy, a Las Vegas utility (Galluci 2023). This could power 2,600 houses at one time, a small contribution, but this is better than forty other geothermal systems in the world, said Fervo. They also claimed it is a continuous source of clean energy that uses minimal land surface area, as compared with wind and solar.

Fig. 19-4. The Blue Mountain Geothermal Power Plant near Winnemucca, Nevada. (Reproduced by permission from Fervo Energy and Google; Galluci 2023.)

At the Project Red site, two wells have been drilled down to 7,700 feet and then kicked off to two horizontal wells 3,250 feet in length in hard impermeable rock. The horizontal legs are fracked along their length to create a permeable reservoir. Presumably, water is pumped down one well and across to exit the other well. While underground, the water is heated to 376°F, which eventually produces steam that runs turbines at the surface to produce electricity.

One clever feature of Project Red is that the electrical power output is synchronized to the grid, so it accommodates the daily variations in grid power usage (Two Bit da Vinci 2024). At nighttime, when residential power needs go up, Project Red can increase geothermal power to compensate. This alleviates the need for a gas-fired peaker plant.

The cost of electricity at Project Red is greater than the DOE's target of \$45/MWh by 2035. But costs will fall when Project Red is scaled up to a new 400 MW commercial project in Utah called Cape Station. Fervo has already started drilling the first of one hundred vertical wells in the new project, which could start producing electricity in 2026 and reach full scale in 2028.

The Google connection is interesting. In 2022 Google set a goal for all its offices and data centers to operate carbon-free by 2030. Google has backed the new geothermal Project Red, which is now transporting clean green electricity to the grid in Nevada, where some large Google data centers exist.

Direct Air Capture

CarbonTech is an exciting growth area in materials science that has both economic and environmental benefits. A start-up in Iceland named Carbfix captures and permanently removes CO_2 by imitating a natural process (Hoek 2023). The CO_2 is dissolved in water and interacts with reactive rock formations, such as basalts, to form stable minerals in a permanent and safe carbon sink.

The global storage potential is huge—greater than the emissions from burning all fossil fuels on earth. In addition, it is a technology that can be scaled in ways that are cheap and eco-friendly.

Heat and Electricity from Hydrogen—without Burning It

New Mexico wants hydrogen-to-heat energy from Australia because the country has a new method to convert hydrogen into heat, at 700°C, without combustion, in a process called HERO, which stands for Hydrogen Energy Release Optimizer (Palmer 2023n). A catalyst was found that encourages hydrogen gas to combine with oxygen gas to form water, and the reaction releases a large amount of heat. An Australian company called Star Scientific has announced it will start building a factory in Albuquerque, New Mexico, in 2024. Initial investment is about $100 million.

Swapping a coal-fired boiler within an existing power plant by a HERO process would cut emissions, according to Star Scientific, and probably cost less. If the inflow to HERO is green hydrogen, there will be no carbon emissions in the entire process.

The efficiency of HERO will be critical for its applications. Once HERO's performance, cost, and durability are demonstrated, it should be able to provide power or heat in many sectors such as long-haul shipping or in heavy industry, where intense heat is required that today comes from fossil fuels. Other sectors include transportation and in commercial and residential buildings. Or the heat can be converted into electricity for just about any electrical application.

Biomass Carbon Removal and Storage

Lawrence Livermore National Laboratory reported a study at the end of 2023 that shows how one-fifth of carbon emissions from the US could be sequestered (Ma 2023). The lab looked at two general approaches: DAC and biomass carbon removal and storage. They argue that 0.5 Gt of emissions could be saved from each approach. The technology exists but not the funding. The cost would be $130 billion to pay for the infrastructure to reach the needed capacity. The main question is durability, since biomass from human waste disposal or forest trash would need to be converted into wood products, hydrogen, biogas, or coke, for example.

Batteries Made of Bricks

Factories that need intense heat, such as those that make steel and cement, may be able to get their power from "heat batteries" developed by startup Rondo Energy. An energy company called EDP has agreed to provide up to 2 GW across Europe starting in 2025. Clay bricks are heated to 1,500°C by renewable energy using elements like those found in a kitchen toaster, and one brick can hold as much energy as a Tesla Model X (Baker 2023).

Takeaways

Two bookends of the question *How far should oil and gas companies pivot toward climate change?* have been presented by two experts who have extensive experience in both oil and climate.

Rystad Energy is confident that net-zero is achievable by 2050. But it appears that Rystad's optimism needs to be tempered by impractical factors that involve CCS and hydrogen.

Pielke on the other hand offers a startling equation: First, fossil fuel emissions would have to be reduced to zero by 2050. Second, one nuclear power plant needs to be constructed per day—from today until 2050.

CCS might be able to save a small fraction of fossil fuels (possibly 5%), but to expand its potential emission removal fraction from to 20% or 50% is completely implausible.

Expected energy transitions such as grid electricity and EVs will hurt the bottom line of companies that are focused on oil and gas.

The IEA says the world has a good chance of tripling renewable energy capacity by 2030, a goal that was set at COP28.

BP shows, in a future scenario that might be the most likely one, a *drastic* reduction in fossil energies from 80% of total energy supply now to 28% by 2050. This sits in contrast to current enthusiasm in the US oil and gas industry. Side by side at 2050, fossil fuels would be at 28%, with renewables at 58%—close to double the fossil fuels.

Following are notes that expand on the future of a few exciting new technologies that may greatly assist the energy transition.

Conventional big-lab nuclear fusion won't help the world reach net-zero by 2050. But small-scale collisions between two high-energy plasma clouds might get there.

The DOE has funded nuclear fission in a small-scale reactor called SMR for many years. But the commercial application faces costs that are too high in both the US and in Australia.

In 2022 Google set a goal for all its offices and data centers to operate carbon-free by 2030. Google, and separately the DOE, have invested in a new geothermal technology based on long horizontal wells fracked many times along their length.

A process called HERO catalyzes hydrogen and oxygen combining to water while producing intense heat. Swapping a coal-fired boiler within an existing power plant by a HERO process would cut emissions and probably cost less.

Factories that need intense heat, such as those that make steel and cement, may be able to get their power from "heat batteries." Clay bricks are heated to 1,500°C by renewable energy. One brick can hold as much energy as a Tesla Model X.

Chapter 20

What Has Been Learned about the Oil and Gas Industry as it Pivots toward Climate Change?

- The Permian basin is the capstone of US oil and gas basins, with its oil production about half, and gas production about a quarter, of total US production in 2023. The basin provides a big part of energy security for the US.

  As an example, ExxonMobil+Pioneer's new resource base of fifteen to twenty years of drilling inventory may become a Permian template for a productive supply chain from upstream drilling to LNG and petrochemical products on the Gulf Coast. Such a level of operations and marketing shores up energy security for the US, and overseas, in a time of uncertainty for the oil and gas industry.

- Vicki Hollub, CEO of Occidental which works the Permian, and President Biden have a lot in common right now: energy security, the energy transition, CCS, and DAC. And, last but not least, attractive contracts and tax credits newly minted by the US government to promote endeavors that reduce GHG emissions.

- In 2022 it became clear to the world that energy security needs to be protected because only oil and gas energies, such as LNG, could fill the gap created by the Russian war in Ukraine. But the dilemma is that oil and gas companies and use of their products cause about 50% of carbon emissions.

  BP has led in diversifying from oil and gas to renewables. BP's stated goal was 40% renewables by 2030, but in 2023 this was changed to 25% by 2030. In 2021, ExxonMobil made a significant pivot toward climate change. These two oil majors are influencing the world in how to integrate climate change into an oil and gas industry.

- To provide energy security, LNG replaced all the EU gas lost from Russia within two years. US export revenues grew exponentially through 2022 in a golden age for LNG. At the end of 2023, Europe took 68% of US exports, while Asia took 19%. Export volumes and wealth from LNG could potentially lead to $100 billion investment in new LNG developments in the US.

- Global warming has been proven, the oil and gas industry is responsible for about 50% of emissions, and Exxon freely and publicly confirmed this forty years ago.

- Regarding CCS, the US could inject into old oil fields its total source of CO_2, roughly 6 billion tpy, for twenty-three years. Or they could store half the US source for forty-six years. There is practically no limit on potential CO_2 storage volume in the US.

  Congress, in 2021 and 2022, upped the 45Q tax credits and boosted direct funding related to CCS and other projects by an enormous amount—almost half a trillion dollars over the next five years from 2023. This has guaranteed a transformative change for CCS in the US.

  Stopping temperatures from rising by 1.5°C–2°C by 2050 would require CCS global capacity to increase by one hundred to two hundred times the present capacity of 40 million tpy. This would represent a 20% growth in CCS year over year. The carbon capture industry would grow as big as the present oil and gas industry by 2050. The costs of CCS projects under these conditions would be passed along to the consumer, and this would imply higher prices at the gas station and steeper electricity costs from coal- or gas-burning power plants.

- ExxonMobil+Pioneer has set on an ambitious course to limit its carbon emissions. They have nominated three buckets to control carbon emissions: reducing carbon emissions in the field, CCS, and hydrogen generation. The company is at the forefront to repeal emissions in all these categories.

  ExxonMobil has plans to build a 1 Bcfd blue hydrogen and ammonia production plant linked with a CCS system at its refinery in Baytown, Texas. The hydrogen would be used for ExxonMobil's olefin production plant, where carbon emissions could be reduced by 30%.

The Baytown project would be ExxonMobil's contribution to a visionary joint venture with other partners to establish a CCS hub in Houston, valued at $100 billion.

- Oxy's new venture is an imposing wall of fans that can suck CO_2 out of the air (DAC). Their vision is one hundred such facilities in the US by 2035 and eventually one thousand franchised operations across the world.

  Warren Buffet may see business opportunity for Oxy's DAC at the front end of CCS growth in the US. This may be a safe investment in the short term, but in the long term the CCS business looks to be expensive and impractical.

- BP's new ventures in Pilbara, Peacock, and TravelCenters of America point to its diversification. BP's budget for transition energy growth is close to $9 billion per year through 2030. Half of this will go into convenience stores, bioenergy, EV charging, and later hydrogen.

  AREH in the Pilbara iron ore area of Australia may become one of the largest green hydrogen hubs in the world. Markets in local Australia as well as bustling cities in nearby Southeast Asia will enable AREH and help BP to reach its goal of capturing 10% of global hydrogen markets.

- In BP's scenarios for the future, the Russian war in Ukraine is likely to swing the pendulum away from "business as usual" and toward the Paris goal of net-zero emissions by 2050.

  In contrast to the present enthusiasm in the US oil and gas industry, BP forecasts a drastic reduction in fossil energies from 80% of total energy consumption now to somewhere between 28% and 55% by 2050. Coal will fall the most, oil will fall, and natural gas may or may not fall because of an ever-rising demand for LNG around the world.

- Large and small companies are jumping on hydrogen, including ExxonMobil, BP, Shell, Chevron, TotalEnergies, NEOM in Saudi Arabia, and ABEL in Australia.

  But there are serious headwinds. Reputable projections show hydrogen usage would likely rise to a maximum of 15% of total energy consumption. By one account, $7 trillion or more would be the cost to make only 5% hydrogen. Hydrogen production is unlikely to replace oil and gas production for cars and gas-fired power plants.

Broad market application of hydrogen—beyond a niche market of aviation/shipping fuel and the manufacture of cement, steel, and chemicals (fertilizers and plastics)—is uncertain and may be a case of too little, too late, and too expensive.

- President Biden's goal is that new sales of EVs in the US will be 50% of all car sales by 2030. This suggests a 25% decline in gasoline cars by 2030, implying a 17% drop in crude oil production by 2030.

  If in the US renewables provide 80% of total electricity by 2030, per Biden's goal, then production of natural gas in the US would fall by 28% by 2030.

  The actual EV numbers for 2023 are: Hybrid cars are up to 8% of new car sales. Pure-electric cars are up to 7%, and plug-in hybrids are at 2%. Altogether, these are close to 16% but still a long way from non-electric cars, at 84%. Elon Musk said his goal is to make twenty million electric cars each year by 2030, and this would require building ten or twelve more gigafactories.

  Ford said it will join the Tesla network of US Superchargers in two stages in 2024 and 2025.

  Surging batteries are a proxy for displacing gasoline in cars and trucks and for displacing coal and natural gas in power plants. The oil and gas industry should keep a watchful eye on growth in EV batteries, home batteries for rooftop solar, and grid batteries that store energy from wind and solar farms, because they will become the despoilers of oil and gas production.

- Australia features in the development of battery systems. The 2017 Tesla battery in Hornsdale, Australia, was a revolutionary idea in the search for renewable resilience and, at 100 MW, was the biggest in the world.

  Australia recognized early that a combination of solar, wind, and batteries has deep benefits. Several very large grid batteries are being built on sites where coal-fired power plants are being decommissioned.

  One proactive company, AGL, expects that electrification of cars and trucks will result in a massive shift to wind and solar. AGL also anticipates that big battery storage will make better financial returns on their investment than solar or wind farms.

Averaged over a year, 72% of South Australia's electricity comes from wind and solar and grid batteries. Starting from virtually no renewables, this result was achieved in just over sixteen years—remarkable for a state with such a tiny economy.

A radical plan is to build a "community" virtual power plant that would create a network of fifty thousand rooftop solar systems and Tesla Powerwall home battery systems across South Australia. The mission of Tesla Energy Ventures is to power one's car, home, and entire community with renewable electricity from Tesla.

- In mid-2023, five oil companies—ExxonMobil, Shell, Chevron, ConocoPhillips, and BP—were named in a lawsuit brought by the California attorney general.

  ExxonMobil's contribution to global emissions is enormous because it's the biggest oil major in the US and one of the biggest in the world.

  Global warming is caused by GHG emissions, the majority of which, about 73%, come from fossil fuels.

  It was alleged that Exxon, in public, cast doubt on the science of climate change in the past twenty years. But Exxon's main predictions in the 1970s and 1980s were about global warming, not climate change, which is a different creature.

  A jury in such a trial would need to understand that the evidence is much stronger for direct indicators of climate change, such as Arctic ice melting and glacier retreats, than it is for indirect indicators, such as killer extreme weather events, despite what some of the press and climate alarmists argue.

  Instead of deliberately misleading the public, an alternative explanation is that ExxonMobil merely expressed uncertainty about the more serious effects of climate change—the killer extreme weather events.

- In climate change the dominant dangers are droughts, wildfires, flooding, and hurricanes. If these four killer weather extremes don't show any global worsening trend over the past forty to fifty years, even though the global temperature has risen 1°C in this time, why should the world be so worried about the next 0.3°C rise in global warming to reach the Paris danger number of 1.5°C above preindustrial times?

So the urgency about climate change on a global basis is not as dire as climate alarmists are saying. Also, the oil and gas industry is not responsible for worsening of killer weather extremes. Yet the industry is culpable for direct effects of climate change such as polar icecaps melting and glaciers retreating, because about 50% of greenhouse emissions come from oil and gas.

- Climate change is not responsible for increases in billion-dollar weather extremes, either in the US or in the world. Inconsistencies exist within the UN-based IPCC's position on weather disasters caused by climate change. Although this is confusing, the UN goals to mitigate climate change should be maintained because of their direct effects such as polar icecaps melting and sea level rise.

- It's easier and more effective in the short term to find and patch up methane leaks than it is to decarbonize a power plant or build an EV to replace a gasoline car. Rystad Energy insist the race to reach net-zero by 2050 consists of two parts. First is a short-term sprint to reduce flaring and methane leaks, now 10 Gt/yr, by 2030. Second is a long-term marathon to decarbonize the general economy by removing what's now 40 Gt/yr by 2050.

- The essence of the US pause in permits for new LNG ventures is this. On one hand, LNG provides energy security and reduces emissions by displacing coal-fired power plants. On the other hand, LNG causes carbon emissions from methane leakage and combustion of LNG.

  A rational solution does exist for the leaky problem of LNG, which refers to emissions connected with natural gas production and distribution. The leaks are not difficult to repair, repairs can be done quickly, and the cost is inexpensive. Certified gas is gas that has been certified by an independent third party to have been produced and distributed with lower methane leaks.

  The other problem is that LNG still burns to CO_2 when it's consumed. The emissions problem of LNG, in total, along with estimated cost of climate change events, will have to be rationalized against the need for worldwide energy security. Only then will the US administration be able to decide whether to undo or continue the pause in permitting of new LNG projects.

- Mitigation and adaptation of direct climate change effects is still necessary for the world. It is still urgent to arrest melting of the polar icecaps, retreating glaciers, rising of sea levels, and bleaching of corals. It's urgent not only because these direct effects cause a significant amount of distress and damage in the world, but also because tipping points may occur and accelerate global warming.

- It would be helpful to avoid the alarmist cries that the world will collapse in a few years because of global warming. It is not useful nor correct to point the finger at every wildfire or hurricane and attribute it to global warming when the long-term global data reveal no worsening trend for the killer weather extremes.

  A more reasoned approach would be that the killer weather extremes have not worsened over a 1°C rise in temperature, so they shouldn't worsen over the next 0.3°C rise to the Paris danger level. This at least alleviates the urgency for achieving net-zero by 2050, which many say is unlikely anyway. And it offers a fresh approach for oil and gas people to sit down at the table with climate and environmental experts to discuss issues of energy security and climate security side by side.

- So how far should oil and gas companies pivot toward climate change? Two bookends of the question have been presented. One is a cautious optimist; the other hints that success will come only by new technology, or a moonshot.

  Expected energy transitions such as electricity and EVs will hurt the bottom line of companies that are totally focused on oil and gas.

  The IEA says the world has a good chance of tripling renewable energy capacity by 2030, a goal that was set at COP28.

  BP shows, in a future scenario that could be the most likely one, a *drastic* reduction in fossil energies from 80% of total energy supply now to 28% by 2050. This sits in contrast to current enthusiasm in the US oil and gas industry. Side by side at 2050, fossil fuels will be at 28%, with renewables at 58%—close to double the fossil fuels.

  A few exciting new technologies may greatly assist the energy transition.

  Conventional big-lab nuclear fusion will not. But small-scale fusion collisions between two high-energy plasma clouds might.

Small-scale nuclear fission reactors, called SMR, are on the market but may be too expensive. A medium scale Natrium reactor looks promising.

Google and, separately, the DOE have invested in a new geothermal technology based on long horizontal wells fracked many times along their length. The results are encouraging.

Swapping a coal-fired boiler in an existing power plant by a HERO process would cut emissions and probably cost less. The HERO process produces heat at 700°C by fusing hydrogen and oxygen via a catalyst, without combustion.

Factories that need intense heat, such as those that make steel and cement, may be able to get their power from "heat batteries." Clay bricks are heated to 1,500°C by renewable energy. One brick can hold as much energy as a Tesla Model X.

Bibliography

2023 Energy Outlook. 2023. BP. Updated July 2023. https://www.bp.com/content/dam/bp/business-sites/en/global/corporate/pdfs/energy-economics/energy-outlook/bp-energy-outlook-2023.pdf.

2023 Advancing Climate Solutions Report. 2023. ExxonMobil. chrome-extension://efaidnbmnnnibpcajpcglclefindmkaj/https://corporate.exxonmobil.com/-/media/global/files/advancing-climate-solutions-progress-report/2023/2023-advancing-climate-solutions-progress-report.pdf.

2024 Advancing Climate Solutions Report. 2024. ExxonMobil. https://corporate.exxonmobil.com/sustainability-and-reports/advancing-climate-solutions.

Addison, Velda. 2024. "Occidental Reports 'Great Progress' on Net-zero Pathway, DAC." Hart Energy, February 21, 2024. https://www.hartenergy.com/exclusives/occidental-reports-great-progress-net-zero-pathway-dac-208248.

Ainger, John, Jess Shankleman, and Jennifer A. Dlouhy. 2023. "COP28 Nations Reach First-Ever Deal to Move Away from Fossil Fuels." Bloomberg. Updated December 13, 2023. https://www.bloomberg.com/news/articles/2023-12-13/cop28-ends-with-deal-on-transition-away-from-fossil-fuels?srnd=green-united-nations-climate-summit.

Alimonti, Gianluca, Luigi Mariani, Franco Prodi, and Renato Angelo Ricci. 2022. "A Critical Assessment Of Extreme Events Trends in Times of Global Warming." *The European Physical Journal Plus* 137, No. 112. https://doi.org/10.1140/epjp/s13360-021-02243-9.

Angueira, Gabriela Aoun. 2023. "BP's $100 Million Tesla Deal Pushes Big Oil Further into the EV Space." Grist, November 1, 2023. https://grist.org/transportation/bp-tesla-ev-charging/.

"Antarctic Ice and Rising Sea Levels." n.d. Antarctic and Southern Ocean Coalition. Accessed February 9, 2024. https://www.asoc.org/learn/antarctic-ice-and-rising-sea-levels/.

API News. 2023. "New Study: U.S. LNG Critical to Addressing European Supply Gap." American Petroleum Institute. Accessed April 3, 2024. https://www.api.org/news-policy-and-issues/news/2023/12/14/new-study-us-lng-critical-to-addressing-european-supply-gap.

"August 2016 Record Flooding." n.d. National Weather Service. Accessed February 9, 2024. https://www.weather.gov/lix/August2016flood.

"Australia's Grid-Scale Battery Storage Set for Quadruple Growth." 2022. EcoGeneration, July 22, 2022. https://www.ecogeneration.com.au/australias-grid-scale-battery-storage-set-for-quadruple-growth/.

Baddour, Dylan, and Delger Erdenesanaa. 2023. "After Explosion, Freeport LNG Rejoins the Gulf Coast Energy Export Boom." Inside Climate News, February 23, 2023. https://insideclimatenews.org/news/23022023/freeport-lng-texas-louisiana-explosion/.

Baer, Justin, and Dawn Lim. 2021. "The Hedge-Fund Manager Who Did Battle with Exxon—and Won." *Wall Street Journal Online*, June 12, 2021. https://www.wsj.com/articles/the-hedge-fund-manager-who-did-battle-with-exxonand-won-11623470420.

Baker, David R. 2023. "Fighting Climate Change with Hot Bricks, Factory by Factory." Bloomberg, March 6, 2023. https://www.bloomberg.com/news/articles/2023-03-06/fighting-climate-change-with-hot-bricks-factory-by-factory.

Bamber, Jonathan. 2022. "The Arctic is Warming Nearly Four Times Faster than the Rest of the World." PBS News Hour, August 15, 2022. https://www.pbs.org/newshour/politics/the-arctic-is-warming-nearly-four-times-faster-than-the-rest-of-the-world.

Banerjee, Neela, Lisa Song, and David Hasemyer. 2015. "Exxon's Own Research Confirmed Fossil Fuels' Role in Global Warming Decades Ago." Inside Climate News, September 16, 2015. https://insideclimatenews.org/news/16092015/exxons-own-research-confirmed-fossil-fuels-role-in-global-warming/.

"Baytown Operations." n.d. ExxonMobil. Accessed February 7, 2024. https://corporate.exxonmobil.com/locations/united-states/baytown-operations.

Blackmon, David. 2022. "Exxon to Mount a Major Hydrogen, CCS Project at Its Baytown Complex." Forbes, March 2, 2022.

https://www.forbes.com/sites/davidblackmon/2022/03/02/exxon-to-mount-a-major-hydrogen-ccs-project-at-its-baytown-complex/?sh=3ee8482e17ee.

Bousso, Ron, Shadia Nasralla, and Sarah Mcfarlane. 2023. "Insight: Inside BP's Plan to Reset Renewables as Oil And Gas Boom." Reuters, March 7, 2023. https://www.reuters.com/business/sustainable-business/inside-bps-plan-reset-renewables-oil-gas-boom-2023-03-07/.

"BP Begins Construction on New Texas Solar Project." 2023. BP, September 26, 2023. https://www.bp.com/en/global/corporate/news-and-insights/press-releases/bp-begins-construction-on-new-texas-solar-project.html.

BP Energy Outlook: 2023 Edition. 2023. BP. Updated July 5, 2023. https://www.bp.com/en/global/corporate/energy-economics/energy-outlook.html.

"BP Plans Major Green Hydrogen Project In Teesside." 2021. BP, November 29, 2021. https://www.bp.com/en/global/corporate/news-and-insights/press-releases/bp-plans-major-green-hydrogen-project-in-teesside.html.

Brelsford, Robert. 2023. "ExxonMobil Using Integrated Approach on Path to Decarbonize Downstream." *Oil & Gas Journal Online*, May 1, 2023. https://www.ogj.com/energy-transition/article/14293451/exxonmobil-using-integrated-approach-on-path-to-decarbonize-downstream.

Carroll, David. 2023. "South Australia Moves to Fast-Track Big Battery Developments." *PV Magazine Australia Online*, November 2, 2023. https://www.pv-magazine-australia.com/2023/11/02/south-australia-moves-to-fast-track-big-battery-developments/.

Casey, Jessica. 2023. "Wood Mackenzie: Global LNG Market Needs 90 Million tpy of Additional Supply by 2035 to Meet Booming Demand." LNG Industry, November 15, 2023. https://www.lngindustry.com/special-reports/15112023/wood-mackenzie-global-lng-market-needs-90-million-tpy-of-additional-supply-by-2035-to-meet-booming-demand/.

Casey, Tina. 2023. "Shell Makes Another Green Hydrogen Move, Exxon Gets the Blues." Clean Technica. Accessed February 8, 2024. https://cleantechnica.com/2023/03/09/shell-makes-another-green-hydrogen-move-exxon-gets-the-blues/.

Chetwynd, Gareth. 2024. "Billionaire-Owned Adani Flicks On-Switch at Planned 'Largest in World' 30GW India Green Energy Park." Recharge

News, February 14, 2024. https://www.rechargenews.com/energy-transition/billionaire-owned-adani-flicks-on-switch-at-planned-largest-in-world-30gw-india-green-energy-park/2-1-1598305.

Chow, Emily. 2023. "LNG Interest from Asia's Emerging Gas Markets Rises as Prices Ease." Reuters, February 1, 2023. https://www.reuters.com/business/energy/easing-prices-revive-lng-interest-asias-emerging-gas-markets-2023-02-01/.

Chu, Jennifer. 2023. "Explained: The 1.5 C Climate Benchmark." MIT News, August 27, 2023. https://news.mit.edu/2023/explained-climate-benchmark-rising-temperatures-0827.

Clark, Aaron. 2023. "Methane From Oil and Gas Are Worse Than Reported to UN, Satellites Show." Bloomberg, September 14, 2023. https://www.bloomberg.com/news/articles/2023-09-14/satellites-expose-holes-in-global-rules-for-methane-reporting?embedded-checkout=true.

Clark, Lesley. 2023. "Exxon Relied on Hacked Info in Fight against Climate Probes—Court Docs." E&E News, October 16, 2023. https://www.eenews.net/articles/exxon-relied-on-hacked-info-in-fight-against-climate-probes-court-docs/.

Coifman, Jon. 2021. "Study: Cutting Methane Emissions Quickly Could Slow Climate Warming Rate by 30%." Environmental Defense Fund, April 27, 2021. https://www.edf.org/media/study-cutting-methane-emissions-quickly-could-slow-climate-warming-rate-30.

Colthorpe, Andy. 2023. "Grid-Scale Battery Boom as US Quarterly Installs Go Up 32%, Pipeline Grows 45% Year-On-Year." Energy Storage News, August 8, 2023. https://www.energy-storage.news/grid-scale-battery-boom-as-us-quarterly-installs-go-up-32-pipeline-grows-45-year-on-year/.

"Copernicus: Extreme Wildfires in Europe, South America and North America in 2022." 2022. Copernicus, December 12, 2022. https://atmosphere.copernicus.eu/copernicus-extreme-wildfires-europe-south-america-and-north-america-2022.

Crowley, Kevin, David Wethe, and Mitchell Ferman. 2023. "Permian's Slow Burn to Iran-Sized Growth Underlines OPEC's Grip." BNN Bloomberg, April 18, 2023. https://www.bnnbloomberg.ca/permian-s-slow-burn-to-iran-sized-growth-underlines-opec-s-grip-1.1909092.

Damian, Carrington. 2023. "Climate-Heating Gases Reach Record Highs, UN Reports." *The Guardian Online*, November 15, 2023.

https://www.theguardian.com/environment/2023/nov/15/climate-heating-gases-reach-record-highs-un-reports?.

Daniels, Jarad. 2024. "Scaling up through 2030," Global CCS Institute, 2024. https://www.globalccsinstitute.com/wp-content/uploads/2024/01/Global-Status-of-CCS-Report-1.pdf

"Denmark to Build Hydrogen Islands; H2 Green Steel Raises Financing." 2024. Reuters Events, January 24, 2024. https://www.reutersevents.com/renewables/renewables/denmark-build-hydrogen-islands-h2-green-steel-raises-financing.

Dloughy, Jennifer A., and Keith Laing. 2024. "Biden's Newest Rule on Auto Emissions Set to Boost Electric Car Sales," Bloomberg Green, March 20, 2024. https://www.bloomberg.com/news/articles/2024-03-20/biden-set-to-boost-electric-car-sales-with-new-epa-rule.

Douglas, Leah. 2023. "COP28 Summit: Global Dairy Companies Join Alliance to Cut Methane." Reuters, December 5, 2023. https://www.reuters.com/sustainability/climate-energy/global-dairy-companies-announce-alliance-cut-methane-cop28-2023-12-05/.

Drane, Amanda. 2023. "Cheniere Plans Major Expansion at Sabine Pass, Already the Nation's Largest LNG Facility." *Houston Chronicle Online*. Updated February 23, 2023. https://www.houstonchronicle.com/business/energy/article/cheniere-sabine-pass-lng-earnings-17802008.php.

Drane, Amanda. 2024. "The Carbon Storage Race is Underway in East Texas as Big Oil Looks to Cash in on Climate Incentives." *Houston Chronicle Online*, March 10, 2024. https://www.houstonchronicle.com/business/energy/article/carbon-storage-east-texas-18676152.php.

Ebbs, Stephanie, Julia Jacobo, Daniel Manzo, and Daniel Peck. 2023. "Why It Matters That Earth Is on the Brink of 1.5 Degrees Celsius in Warming." ABC News, November 1, 2023. https://abcnews.go.com/US/matters-earth-brink-15-degrees-celsius-warming/story?id=104508627.

Edser, Nick, Daniel Thomas, and Noor Nanji. 2023. "BP Scales Back Climate Targets as Profits Hit Record." BBC News, February 7, 2023. https://www.bbc.com/news/business-64544110.

EIA (US Energy Information Administration). 2022. "U.S. Liquefied Natural Gas Exports to Europe Increased During the First 4 Months of 2022." EIA.gov, June 7, 2022. https://www.eia.gov/todayinenergy/detail.php?id=52659.

EIA (US Energy Information Administration). 2023a. "Natural Gas Weekly Update." EIA.gov, October 25, 2023. https://www.eia.gov/naturalgas/weekly/archivenew_ngwu/2023/10_26/.

EIA (US Energy Information Administration). 2023b. "Europe Was the Main Destination for U.S. LNG Exports in 2022." EIA.gov, March 22, 2023. https://www.eia.gov/todayinenergy/detail.php?id=55920#.

EIA (US Energy Information Administration). 2023c. "U.S. Energy Facts Explained." EIA.gov. Updated August 16, 2023. https://www.eia.gov/energyexplained/us-energy-facts/.

EIA (US Energy Information Administration). 2023d. "LNG-Exporting Gulf Coast States Drove U.S. Natural Gas Demand Growth." EIA.gov, June 27, 2023. https://www.eia.gov/todayinenergy/detail.php?id=56940.

EIA (US Energy Information Administration). 2024a. *Eagle Ford Region Drilling Productivity Report*. EIA.gov. Accessed February 6, 2024. https://www.eia.gov/petroleum/drilling/.

EIA (US Energy Information Administration). 2024b. "U.S. Battery Storage Capacity Expected to Nearly Double in 2024." EIA.gov. Updated January 9, 2024. https://www.eia.gov/todayinenergy/detail.php?id=61202.

EIA (US Energy Information Administration). 2024c. "The United States was the World's Largest Liquefied Natural Gas Exporter in 2023." EIA.gov, April 1, 2024. https://www.eia.gov/todayinenergy/detail.php?id=61683.

Ellacott, Tom, Robert Clarke, Fraser McKay, David Clark, and Matt Woodson. 2023. "The Making of a Megamajor—Exxonmobil Acquires Pioneer Natural Resources." Wood Mackenzie, October 12, 2023. https://www.woodmac.com/news/opinion/the-making-of-a-megamajor-exxonmobil-pioneer/.

EPA (Environmental Protection Agency). 2024. "Overview of Greenhouse Gases." EPA.gov. Updated February 16, 2024. https://www.epa.gov/ghgemissions/overview-greenhouse-gases#methane.

"ExxonMobil Announces Merger with Pioneer Natural Resources in an All-Stock Transaction." 2023. ExxonMobil News Releases, October 11, 2023. https://corporate.exxonmobil.com/news/news-releases/2023/1011_exxonmobil-announces-merger-with-pioneer-natural-resources-in-an-all-stock-transaction.

Farrer, Giles, Murray Douglas, and Massimo Di Odoardo. 2023. "Stick or Twist: Should Gas Resource Holders Target LNG Exports or Blue Ammonia?" Wood Mackenzie. Accessed February 8, 2024. https://www.woodmac.com/horizons/should-gas-resource-holders-target-lng-exports-or-blue-ammonia/.

Fitzpatrick, Tim. 2024. "In Climate Setback, Rocky Mountain Now Says It Plans to Burn Coal in Utah until 2042." *The Salt Lake Tribune Online*, April 1, 2024. https://www.sltrib.com/news/environment/2024/04/01/climate-setback-rocky-mountain-now/.

Forge Utah. 2021. "Drilling into the Geothermal Future: The Landmark Success of the First Deep Deviated Well." YouTube video, posted October 5, 2021. https://www.youtube.com/watch?v=9nM73_5P6z4.

Frazin, Rachel. 2023. "Exxon Makes Foray into Electric Vehicle Battery Minerals." *The Hill Online*, November 13, 2023. https://thehill.com/policy/energy-environment/4307869-exxon-electric-vehicle-battery-lithium/.

Galluci, Maria. 2023. "America's First 'Enhanced' Geothermal Plant Just Got Up and Running." Canary Media, November 28, 2023. https://www.canarymedia.com/articles/geothermal/americas-first-enhanced-geothermal-plant-just-got-up-and-running.

Gambrell, Jon, Peter Prengaman, and Seth Borenstein. 2023. "At COP28 Meeting, Oil Companies Pledge to Combat Methane. Environmentalists Call it a 'Smokescreen.'" AP News. Updated December 2, 2023. https://apnews.com/article/climate-changemethanecop28-58c756d843fbc696b28a330295505be7.

"Gas & Low Carbon Energy." n.d. BP. Accessed February 7, 2024. https://www.bp.com/en_us/united-states/home/what-we-do/gas-and-low-carbon-energy.html.

Gaswirth, Stephanie B., Katherine L. French, Janet K. Pitman, Kristen R. Marra, Tracey J. Mercier, Heidi M. Leathers-Miller, Christopher J. Schenk, Marilyn E. Tennyson, Cheryl A. Woodall, Michael E.

Brownfield, Thomas M. Finn, and Phuong A. Le. 2018. *Assessment of Undiscovered Continuous Oil and Gas Resources in the Wolfcamp Shale and Bone Spring Formation of the Delaware Basin, Permian Basin Province, New Mexico and Texas, 2018.* US Geological Survey Fact Sheet 2018–3073. https://doi.org/10.3133/fs20183073.

Gates, Bill. 2021. *How to Avoid a Climate Disaster: The Solutions We Have and the Breakthroughs We Need.* New York City, NY: Random House.

Gordon, Meghan, and Takeo Kamagai. 2023. "G7 Interview: US Sees Need for New Gas Investments in Light of Russia War, Tight Market." Edited by Shashwat Pradhan. S&P Global Commodity Insights. April 16, 2023. https://www.spglobal.com/commodityinsights/en/market-insights/latest-news/natural-gas/041623-g7-interview-us-sees-need-for-new-gas-investments-in-light-of-russia-war-tight-market.

Graham, Paul, Jenny Hayward, and James Foster. 2023. *GenCost 2023-24: Consultation Draft.* Australia: Commonwealth Scientific and Industrial Research Organisation. https://doi.org/10.25919/xxd6-qz89.

Grinsted, Aslak, Peter Ditlevsen, and Jens Hesselbjerg Christensen. 2019. "Normalized US Hurricane Damage Estimates Using Area of Total Destruction, 1900– 2018." *Proceedings of the National Academy of Sciences* 116, No. 48: 23942–23946. https://doi.org/10.1073/pnas.1912277116.

Guzman, Bryan. 2023. "A 'Brief' History of Carbon Sequestration in the US: What Does the Future Hold?" NSI Tech Live Discussion, streamed August 16, 2023. Accessed February 7, 2024. https://vimeo.com/855097274?share=copy.

"H2 Green Steel Raises More than €4 Billion in Debt Financing for the World's First Large-Scale Green Steel Plant." 2024. H2 Green Steel, January 22, 2024. https://www.h2greensteel.com/latestnews/h2-green-steel-raises-more-than-4-billion-in-debt-financing-for-the-worlds-first-large-scale-green-steel-plant.

Hart Energy Staff. 2024. "Texas LNG Export Terminal Completes Required Permitting for FID." Hart Energy, January 26, 2024. https://www.hartenergy.com/exclusives/texas-lng-export-terminal-completes-required-permitting-fid-207937.

Hausfather, Zeke. 2021. "Global CO2 emissions have been flat for a decade, new data reveals." Carbon Brief, April 11, 2021. https://www.carbonbrief.org/global-co2-emissions-have-been-flat-for-a-decade-new-data-reveals/.

Hoek, Marga. 2023. *Tech For Good: Imagine Solving the World's Greatest Challenges*. Milton Park, Abingdon, Oxfordshire: Routledge.

Howarth, Robert W. 2023. "The Greenhouse Gas Footprint of Liquefied Natural Gas (LNG) Exported from the United States." In review at a peer-reviewed Journal. Accessed February 16, 2024. https://www.research.howarthlab.org/publications/Howarth_LNG_assessment_preprint_archived_2023-1103.pdf.

"Hydrogen Offtake is Tiny but Growing." 2023. BloombergNEF, November 14, 2023. https://about.bnef.com/blog/hydrogen-offtake-is-tiny-but-growing/.

IEA (International Energy Agency). n.d. "CCUS in the Transition to Net-Zero Emissions." IEA.org. Accessed March 16, 2024. https://www.iea.org/reports/ccus-in-clean-energy-transitions/ccus-in-the-transition-to-net-zero-emissions.

IEA (International Energy Agency). 2023. "Executive Summary." In *The Oil and Gas Industry in Net Zero Transitions*. Accessed February 9, 2024. https://www.iea.org/reports/the-oil-and-gas-industry-in-net-zero-transitions/executive-summary.

IEA (International Energy Agency). 2024. "Massive Expansion of Renewable Power Opens Door to Achieving Global Tripling Goal Set at COP28." In *Renewables 2023*. Accessed February 9, 2024. https://www.iea.org/news/massive-expansion-of-renewable-power-opens-door-to-achieving-global-tripling-goal-set-at-cop28.

International Monetary Fund. n.d. "Global Price of LNG, Asia." FRED, Federal Reserve Bank of St. Louis. Last updated on March 16, 2024. https://fred.stlouisfed.org/series/PNGASJPUSDM.

IPCC (Intergovernmental Panel on Climate Change). 2022. *Climate Change 2022: Impacts, Adaptation, and Vulnerability.* Edited by H.-O. Pörtner, D.C. Roberts, M. Tignor, E.S. Poloczanska, K. Mintenbeck, A. Alegría, M. Craig, S. Langsdorf, S. Löschke, V. Möller, A. and Okem, B. Rama. Sixth Assessment Report. https://doi.org/10.1017/9781009325844.

Jacobs, Nicole. 2023. "Report: American Oil and Natural Gas Industry Responsible for 10+ Million Jobs." Energy In Depth, May 16, 2023. https://www.energyindepth.org/report-american-oil-and-natural-gas-industry-responsible-for-10-million-jobs/.

Jacobs, Trent. 2022. "Oxy CEO Vicki Hollub Reflects on Career, Climate, and Carbon." *Journal of Petroleum Technology Online*, February 7, 2022.

https://jpt.spe.org/oxy-ceo-vicki-hollub-reflects-on-career-climate-and-carbon.

Jacobs, Trent. 2023a. "Oxy Doubles Down on Low-Carbon Push Despite Delay With Direct Air Capture." *Journal of Petroleum Technology Online*, March 1, 2023. https://jpt.spe.org/oxy-doubles-down-on-low-carbon-push-despite-delay-with-direct-air-capture.

Jacobs, Trent. 2023b. "ExxonMobil to Become True Permian Juggernaut With Pioneer Acquisition." *Journal of Petroleum Technology Online*, October 11, 2023. https://jpt.spe.org/exxonmobil-to-become-true-permian-juggernaut-with-pioneer-acquisition.

Jacobs, Trent. 2023c. "Electric Rush: ExxonMobil To Drill Enough Lithium Wells To Make 1 Million EVs a Year." *Journal of Petroleum Technology Online*, November 13, 2023. https://jpt.spe.org/electric-rush-exxonmobil-to-drill-enough-lithium-wells-to-make-1-million-evs-a-year.

Johnson, Lamar. 2024. "LG Energy Unveils 10 Grid-Scale Battery Storage Projects for 2024." Utility Dive, January 8, 2024. https://www.utilitydive.com/news/lg-energy-unveils-10-grid-scale-battery-storage-projects-2024/703900/.

Kahn, Brian. 2024. "CES Highlights the Climate-Centric Future of Consumer Tech." Bloomberg, January 11, 2024. https://www.bloomberg.com/news/articles/2024-01-11/ces-in-las-vegas-shows-climate-friendly-tech-is-going-mainstream?embedded-checkout=true.

Kerr, Sonia, Kevin Baxter, Vivien Lebbon, Alishia Markwell, Mark Thomton, and Hla Myat Mon. 2023. "Australia Leads Global Market for Battery Energy Storage Systems." Wood Mackenzie, July 17, 2023. https://www.woodmac.com/press-releases/australia-leads-global-market-for-battery-energy-storage-systems/.

King, Ben. 2023. "Why Are BP, Shell, and Other Oil Giants Making So Much Money Right Now?" BBC, February 11, 2023. https://www.bbc.com/news/business-64583982.

Koonin, Steven E. 2021. *Unsettled: What Climate Science Tells Us, What It Doesn't, and Why It Matters*. Dallas, TX: BenBella Books.

Kumar, Arunima. 2020. "Pioneer Natural Sets Targets to Reduce Greenhouse Gas, Methane Emissions." Edited by Amy Caren Daniel. Reuters, December 2, 2020. https://www.reuters.com/article/us-pioneer-natl-rsc-carbon/pioneer-natural-sets-targets-to-reduce-greenhouse-gas-methane-emissions-idUSKBN28C3DQ/.

Kusnetz, Nicholas. 2023. "Exxon Accurately Predicted Global Warming, Years Before Casting Doubt on Climate Science." Inside Climate News, January 12, 2023. https://insideclimatenews.org/news/12012023/exxon-doubt-climate-science/.

Lavelle, Marianne. 2019. "Former Exxon Scientists Tell Congress of Oil Giant's Climate Research Before Exxon Turned to Denial." Inside Climates News, October 24, 2019. https://insideclimatenews.org/news/24102019/exxon-scientists-climate-research-testify-congess-denial/.

"Leading the Green Economy." n.d. Government of South Australia. Accessed February 8, 2024. https://www.energymining.sa.gov.au/industry/modern-energy/leading-the-green-economy.

Lee, Sheryl Tian Tong, Selina Xu, and Isabel Kua. 2023. "Singapore Is Fighting Rising Seas to Save $50 Billion in Real Estate." Bloomberg, October 9, 2023. https://www.bloomberg.com/news/features/2023-10-09/singapore-is-saving-50-billion-in-real-estate-from-rising-seas-climate-change.

Lewis, Michelle. 2023. "The World's Largest Battery Storage System Just Got Even Larger." Electrek, August 3, 2023. https://electrek.co/2023/08/03/worlds-largest-battery-storage-system-just-got-even-larger/.

Liao, Ruth. 2023. "NextDecade Will Move Ahead with $18.4 Billion Texas LNG Project." BNN Bloomberg, July 12, 2023. https://www.bnnbloomberg.ca/nextdecade-will-move-ahead-with-18-4-billion-texas-lng-project-1.1944934.

Lindner, Jannik. 2023. "Tesla Car Fire Statistics and Trends in 2024." Gitnux. Updated December 23, 2023. https://gitnux.org/tesla-car-fire-statistics/.

L., Jennifer. 2023. "Occidental Petroleum Quietly Abandons Biggest Carbon Capture Plant." CarbonCredits.com. Accessed March 16, 2024. https://carboncredits.com/occidental-petroleum-quietly-abandons-biggest-carbon-capture-plant/.

Ma, Michelle. 2023. "The US Could Remove 1 Billion Tons of Carbon From the Air—for $130 Billion." Bloomberg, December 11, 2023.

https://www.bloomberg.com/news/articles/2023-12-11/united-states-carbon-removal-infrastructure-could-cost-130-billion.

MacDiarmid, Stacy. 2018. "New Study Finds U.S. Oil and Gas Methane Emissions Are 60 Percent Higher Than EPA Reports." Environmental Defense Fund, June 21, 2018. https://www.edf.org/media/new-study-finds-us-oil-and-gas-methane-emissions-are-60-percent-higher-epa-reports-0.

Maguire, Gavin. 2022. "U.S. LNG Exports Both a Lifeline and a Drain for Europe in 2023." Reuters, December 21, 2022. https://www.reuters.com/business/energy/us-lng-exports-both-lifeline-drain-europe-2023-maguire-2022-12-20/.

"Mapping Methane Emissions on a Global Scale." 2020. The European Space Agency, April 5, 2020. https://www.esa.int/Applications/Observing_the_Earth/Copernicus/Sentinel-5P/Mapping_methane_emissions_on_a_global_scale.

Marshall, Aarian. 2024. "Here Comes the Flood of Plug-In Hybrids." WIRED, March 25, 2024. https://www.wired.com/story/here-comes-the-flood-of-plug-in-hybrids/.

Masters, Jeff. 2021. "World Hammered by Record 50 Billion-Dollar Weather Disasters in 2020." Yale Climate Connections, January 25, 2021. https://yaleclimateconnections.org/2021/01/world-hammered-by-record-50-billion-dollar-weather-disasters-in-2020/.

Mathews, Chris. 2023a. "EIA: Permian Basin to Hit Record Oil, Gas Production in May." Hart Energy, April 19, 2023. https://www.hartenergy.com/exclusives/eia-permian-basin-hit-record-oil-gas-production-may-204796.

Mathews, Chris. 2023b. "Exxon Acquiring Pioneer for $60B as Permian Oil Takes Center Stage." Hart Energy, October 11, 2023. https://www.hartenergy.com/exclusives/exxon-acquiring-pioneer-60b-permian-oil-takes-center-stage-206742.

Mathews, Chris. 2023c. "Exxon Bringing Longer Laterals, New Tech to Pioneer Acreage." Hart Energy, October 11, 2023. https://www.hartenergy.com/exclusives/exxon-bringing-longer-laterals-new-tech-pioneer-acreage-206750.

Mathis, Will. 2023. "Solar Power Investment Set to Surpass Oil Production Spending This Year." Bloomberg, May 25, 2023. https://www.bloomberg.com/news/articles/2023-05-25/solar-power-investment-set-to-surpass-oil-production-spending-this-year.

Mathis, Will, and John Ainger. 2024. "The $40 Billion Fund That's Gambling With Europe's Climate Goal." Bloomberg, April 2, 2024. https://www.bloomberg.com/news/articles/2024-04-02/eu-s-40-billion-clean-tech-fund-makes-big-bets-early-stumbles.

Maue, Ryan N. 2024. "Global Tropical Cyclone Activity." Climatlas. Updated February 7, 2024. https://climatlas.com/tropical/.

McCarthy, Dan. 2024. "Chart: Gas-Powered Cars are Losing Market Share to EVs and Hybrids." Canary Media, March 1, 2024. https://www.canarymedia.com/articles/electric-vehicles/chart-gas-powered-cars-are-losing-market-share-to-evs-and-hybrids.

McDonald, Zane. 2023. "COP28 Produces New Climate Pledges from Oil and Gas Companies. We're Leaning into the Tough Next Step: Keeping Them." Medium, December 21, 2023. https://medium.com/catalyst-by-gti/cop28-produces-new-climate-pledges-from-oil-and-gas-companies-489cc66b50a0.

McGee, Matt. 2019. "New Mexico's Methane Problem Worsens as Permian Production Soars." Environmental Defense Fund, April 11, 2019. https://www.edf.org/media/report-new-mexicos-methane-problem-worsens-permian-production-soars.

Menon, Shanti, 2024. "This methane-hunting satellite is built to fight climate change," Vital Signs, EDF, 27 January 2024. https://vitalsigns.edf.org/story/methane-hunting-satellite-built-fight-climate-change.

Meredith, Sam. 2023. "The West's Oil War against Russia is Starting to Lose Momentum." CNBC, May 24, 2023. https://www.cnbc.com/2023/05/24/ukraine-the-wests-oil-war-against-russia-is-losing-momentum.html.

Mihalascu, Dan. 2023. "California Tops US EV Adoption: 25% EV Share of Total Sales in H1 2023." InsideEVs, September 27, 2023. https://insideevs.com/news/688779/california-tops-us-ev-adoption-25-percent-share-total-sales-h1-2023/.

"MiQ Launches World's First Comprehensive Greenhouse Gas Certification and Registry for LNG." 2023. MiQ. Accessed February 9, 2024. https://miq.org/miq-launches-worlds-first-comprehensive-greenhouse-gas-certification-and-registry-for-lng/.

Mohleji, Shalini, and Roger Pielke Jr. 2014. “Reconciliation of Trends in Global and Regional Economic Losses from Weather Events: 1980–2008.” *Natural Hazards Review* 15, No. 4. https://doi.org/10.1061/(ASCE)NH.1527-6996.0000141.

Morris, Gregory. 2024. “Louisiana ‘Ready to Roll’ After EPA Grants Authority Over CCS Wells.” Hart Energy, January 4, 2024. https://www.hartenergy.com/exclusives/louisiana-ready-roll-after-epa-grants-authority-over-ccs-wells-207682.

Morton, Adam, and Lisa Cox. 2022. “‘Devastating’: 91% of Reefs Surveyed on Great Barrier Reef Affected by Coral Bleaching in 2022.” *The Guardian Online*, May 10, 2022. https://www.theguardian.com/environment/2022/may/10/devastating-90-of-reefs-surveyed-on-great-barrier-reef-affected-by-coral-bleaching-in-2022.

Murtaugh, Dan. 2023. “What a Chinese Road Trip Taught Us About EVs and the Future of Oil.” Bloomberg, December 22, 2023. https://www.bloomberg.com/news/newsletters/2023-12-22/ev-road-trip-what-a-journey-across-china-taught-us-about-the-future-of-oil.

“NEOM Accelerates Progress Towards Green Hydrogen Future.” 2023. NEOM, May 24, 2023. https://www.neom.com/en-us/newsroom/neom-accelerates-green-hydrogen-future#:~:text=Located%20in%20Oxagon%2C%20the%20NEOM,tonnes%20of%20CO2%20annually.

N. R., Sethuraman, and Forrest Crellin. 2023. “TotalEnergies to Invest $300 Million in Joint Venture with Adani Green Energy.” Reuters, September 20, 2023. https://www.reuters.com/business/energy/frances-total-invest-300-mln-form-jv-with-adani-green-2023-09-20/.

Nangoy, Fransiska, and Bernadette Christina. 2023. “Indonesia Says Exxon Mobil Plans to Invest Up to $15 bln in the Country.” Edited by Muralikumar Anantharaman and Tom Hogue. Reuters, November 15, 2023. https://www.reuters.com/business/energy/indonesia-says-exxon-mobil-plans-invest-up-15-bln-country-2023-11-16/.

Neo. “Why Saudi Arabia is Building a Floating City.” 2022. YouTube video, posted June 30, 2022. https://www.youtube.com/watch?v=L0Le8bT1RLE.

Net Zero Ambition Progress Update. 2023. BP, March 2023.

https://www.bp.com/en/global/corporate/sustainability/reporting-centre-and-archive.html#reports.

Newman, Rick. 2023. "Biden Finally Glimpses the Importance of Oil." Yahoo News. Updated March 13, 2023. https://news.yahoo.com/biden-finally-glimpses-the-importance-of-oil-145829483.html?guccounter=1.

Nichols, Sam, and Jesse Hyland. 2024. "Two Killed in NSW House Blaze Believed to be State's First Recorded Lithium-Ion Battery-Related Deaths, Fire and Rescue NSW Says." ABC News, March 4, 2024. https://www.abc.net.au/news/2024-03-05/first-deaths-lithium-ion-batteries-nsw-teralba/103548972.

NOAA (National Oceanic and Atmospheric Administration.) 2024a. "Monthly Global Climate Report for Annual 2023." National Centers for Environmental Information. January 2024. https://www.ncei.noaa.gov/access/monitoring/monthly-report/global/202313.

NOAA (National Oceanic and Atmospheric Administration.) 2024b. "U.S. Billion-Dollar Weather and Climate Disasters." National Centers for Environmental Information. Updated March 8, 2024. https://doi.org/10.25921/stkw-7w73.

"Note from the CEO—November 2023." 2023. RystadEnergy, November 28, 2023. https://www.rystadenergy.com/insights/note-from-the-ceo-november-2023.

Ocko, Ilissa B., Tianyi Sun, Drew Shindell, Michael Oppenheimer, Alexander N. Hristov, Stephen W. Pacala, Denise L. Mauzerall, Yangyang Xu, and Steven P. Hamburg. 2021. "Acting Rapidly to Deploy Readily Available Methane Mitigation Measures by Sector Can Immediately Slow Global Warming." *Environmental Research Letters* 16, No. 5. https://doi.org/10.1088/1748-9326/abf9c8.

Office of Energy Efficiency & Renewable Energy. 2021. "Should I Get Battery Storage for My Solar Energy System?" Photograph by Christine Bennett. Solar Energy Technologies Office, November 22, 2021. https://www.energy.gov/eere/solar/articles/should-i-get-battery-storage-my-solar-energy-system.

OGJ Editors. 2023. "WoodMac: US Could See $100 Billion in New LNG Projects over Next 5 Years." *Oil & Gas Journal Online*, February 23, 2023.

https://www.ogj.com/general-interest/economics-markets/article/14290197/woodmac-us-could-see-100-billion-in-new-lng-projects-over-next-5-years.

Olano, Maria Virginia. 2023. "Chart: The US is Officially the World's Biggest LNG Exporter." Canary Media, October 20, 2023. https://www.canarymedia.com/articles/liquefied-natural-gas/chart-the-us-is-officially-the-worlds-biggest-lng-exporter.

Olano, Maria Virginia. 2024. "Chart: Is LNG Worse for the Climate than Coal?" Canary Media, January 26, 2024. https://www.canarymedia.com/articles/liquefied-natural-gas/chart-is-lng-worse-for-the-climate-than-coal.

O'Reilly, Callum. 2024. "QatarEnergy's Plans for North Field Shows Massive Scale of Reserves." Hydrocarbon Engineering, February 27, 2024. https://www.hydrocarbonengineering.com/gas-processing/27022024/qatarenergys-plans-for-north-field-shows-massive-scale-of-reserves/.

Osborne, James. 2023. "Biden Wants Hydrogen, Carbon Capture Scaled up within a Decade, but Is the Technology Ready?" *Houston Chronicle Online*. Updated November 6, 2023. https://www.houstonchronicle.com/business/energy/article/clean-hydrogen-carbon-capture-biden-18459766.php.

Oswald, Erik. 2021. "Houston Carbon Capture and Storage Gains Industry Support, Now Needs Policy Support." ExxonMobil, September 23, 2021. https://corporate.exxonmobil.com/what-we-do/delivering-industrial-solutions/carbon-capture-and-storage/houston-ccs-gains-industry-support.

"Our Electricity Supply and Market." n.d. Government of South Australia. Accessed February 8, 2024. https://www.energymining.sa.gov.au/consumers/energy-grid-and-supply/our-electricity-supply-and-market.

Palmer, Ian. 2020a. *The Shale Controversy*. Self-published, Kindle Direct Publishing.

Palmer, Ian. 2020b. "Put It Underground: A Feasible Biden Plan for Fossil Energy Emissions–Except for the Cost." Forbes.com, December 29, 2020.

https://www.forbes.com/sites/ianpalmer/2021/12/29/put-it-underground-a-feasible-biden-plan-for-fossil-energy-emissions--except-for-the-cost/?sh=271aaadb26e8.

Palmer, Ian. 2021a. "As Electricity And Transport Turn 'Green'—Is There Any Future in Gasoline?" Forbes.com, April 13, 2021. https://www.forbes.com/sites/ianpalmer/2021/04/13/greening-of-electrical-and-transport-sectors--what-the-numbers-mean-for-the-future-of-oil-and-gas/?sh=50cad696170e.

Palmer, Ian. 2021b. "Carbon Capture Won't Save Oil and Gas." Barron's, November 22, 2021. https://www.barrons.com/articles/carbon-capture-wont-save-oil-and-gas-51637592327.

Palmer, Ian. 2021c. "Scale Up of CO2 Capture and Storage Looks Massive, Expensive and Impractical to Manage." Forbes.com, November 29, 2021. https://www.forbes.com/sites/ianpalmer/2021/11/29/scale-up-of-carbon-capture-and-storage-looks-massive-expensive-and-impractical-to-manage/?sh=7c991c066b10.

Palmer, Ian. 2021d. "Big Liquid Hydrogen Project in Teesside, UK—BP are Still Pushing the Envelope." Forbes.com, December 17, 2021. https://www.forbes.com/sites/ianpalmer/2021/12/17/big-liquid-hydrogen-project-in-teesside-uk--bp-are-still-pushing-the-envelope/?sh=1c7aafd32359.

Palmer, Ian. 2021e. "EU versus US: Investing in Oil, Gas and Renewables." Hart Energy, July 16, 2021. https://www.hartenergy.com/ep/opinions/eu-versus-us-investing-oil-gas-and-renewables-195164.

Palmer, Ian. 2022a. "The King Of Shale Oil, the Permian Basin, Is Still Riding the Wave." Forbes.com, March 30, 2022. https://www.forbes.com/sites/ianpalmer/2022/03/30/the-king-of-shale-oil-the-permian-basin-is-still-riding-the-wave/?sh=6d02617b6ebf.

Palmer, Ian. 2022b. "The Concept of Scope 3 Greenhouse Gas Emissions, and How to Measure Them for Carbon Management by Fossil Energy and Other Companies. Part 1." Forbes.com, March 24, 2022. https://www.forbes.com/sites/ianpalmer/2022/03/24/the-concept-of-scope-3-greenhouse-gas-emissions-and-how-to-measure-them-for-carbon-management-by-fossil-energy-and-other-companies-part-1/?sh=5c618b5f54d6.

Palmer, Ian. 2022c. "Warren Buffet's Big Money Bet on Occidental Petroleum Company." Forbes.com, July 25, 2022.

https://www.forbes.com/sites/ianpalmer/2022/07/25/warren-buffets-big-money-bet-on-occidental-petroleum-company/?sh=5d65262a43ab.

Palmer, Ian. 2022d. "Energy Security Versus Climate Security—See How Your Country Ranks Among Others—Some Are Predictable, Some Are Surprises." Forbes.com, November 29, 2022. https://www.forbes.com/sites/ianpalmer/2022/11/29/energy-security-versus-climate-security--see-how-your-country-ranks-among-otherssome-are-predictable-some-are-surprises/?sh=195c98b47258.

Palmer, Ian. 2022e. "Energy Inflation, Energy Security, and When Fossil Fuel Is Not Coming to a Dead End." Forbes.com, July 8, 2022. https://www.forbes.com/sites/ianpalmer/2022/07/08/energy-inflation-energy-security-and-when-fossil-fuel-is-not-coming-to-a-dead-end/?sh=7106c6ae497e.

Palmer, Ian. 2022f. "The Surging Hydrogen Economy That Oil and Gas Companies are Tiptoeing Into." Forbes.com, September 29, 2022. https://www.forbes.com/sites/ianpalmer/2022/09/29/the-surging-hydrogen-economy-that-oil-and-gas-companies-are-tiptoeing-into/?sh=37c4603e6341.

Palmer, Ian. 2022g. "Oil and Gas Profits Very High Once Again–What This Feels Like to Energy Consumers." Forbes.com, November 4, 2022. https://www.forbes.com/sites/ianpalmer/2022/11/04/oil-and-gas-profits-very-high-once-again--what-this-feels-like-to-energy-consumers/?sh=505514ae6f36.

Palmer, Ian. 2022h. "Big-Oil Lands on Huge Hydrogen Hub in Australia, but How Viable Is Hydrogen Really." Forbes.com, June 30, 2022. https://www.forbes.com/sites/ianpalmer/2022/06/30/big-oil-lands-on-huge-hydrogen-hub-in-australia-but-how-viable-is-hydrogen-really/?sh=7335fe33385b.

Palmer, Ian. 2022i. "Is Saudi Arabia's New Climate City 'Neom' Future Or Fantasy?" Forbes.com, February 28, 2022. https://www.forbes.com/sites/ianpalmer/2022/02/28/a-new-climate-city-in-a-big-oil-state--saudi-arabia--is-it-future-or-fantasy/?sh=2e502ff16a0a.

Palmer, Ian. 2022j. "Exponential Sales of EVs Means Less Gasoline, Less Crude Oil, Less Greenhouse Gases." Forbes.com, June 18, 2022. https://www.forbes.com/sites/ianpalmer/2022/06/18/exponential-sales-of-evs-means-less-gasoline-less-crude-oil-less-greenhouse-gases/?sh=7e7dda8a6a1d.

Palmer, Ian. 2022k. "Jump in Methane Emissions That Fingers Big Oil—But Wait, Not So Fast." Forbes.com, April 11, 2022. https://www.forbes.com/sites/ianpalmer/2022/04/11/jump-in-methane-emissions-that-fingers-big-oilbut-wait-not-so-fast/?sh=44f04f932a4b.

Palmer, Ian. 2022l. "When Will Nuclear Fusion Put Oil And Gas Out Of Business." Forbes.com, December 18, 2022. https://www.forbes.com/sites/ianpalmer/2022/12/18/when-will-nuclear-fusion-put-oil-and-gas-out-of-business/?sh=6e3852a41439.

Palmer, Ian. 2022m. "Enhanced Geothermal System Uses Oil and Gas Technology to Mine Low-Carbon Energy. Part 2." Forbes.com, May 19, 2022. https://www.forbes.com/sites/ianpalmer/2022/05/19/an-enhanced-geothermal-system-uses-oil-and-gas-technology-to-mine-low-carbon-energy-part-2/?sh=38c96f82b500.

Palmer, Ian. 2023a. "The Golden Age for Liquefied Natural Gas (LNG)." Forbes.com, March 1, 2023. https://www.forbes.com/sites/ianpalmer/2023/03/01/the-golden-age-for-liquefied-natural-gas-lng/?sh=747a416b78c8j.

Palmer, Ian. 2023b. "More on Golden Age of Liquefied Natural Gas." Forbes.com, April 4, 2023. https://www.forbes.com/sites/ianpalmer/2023/04/04/more-on-golden-age-of-liquefied-natural-gas/?sh=3190f5b62f9b.

Palmer, Ian. 2023c. "Put Your Money on the Permian—The Phar Lap of Oil and Gas Basins." Forbes.com, April 25, 2023. https://www.forbes.com/sites/ianpalmer/2023/04/25/put-your-money-on-the-permianthe-phar-lap-of-oil-and-gas-basins/?sh=663514b1dfe1.

Palmer, Ian. 2023d. "Another Mega-Merger, but ExxonMobil Plus Pioneer Will Be the Influence-Shaper for Oil as Well as Climate." Forbes.com, October 23, 2023. https://www.forbes.com/sites/ianpalmer/2023/10/23/another-mega-merger-but-exxonmobil-plus-pioneer-will-be-the-influence-shaper-for-oil-as-well-as-climate/?sh=38e12ce71331.

Palmer, Ian. 2023e. "California Lawsuit: Oil Companies, Global Warming, Climate Change, Deception: What The Jury Needs To Know. Part 1." Forbes.com, September 26, 2023. https://www.forbes.com/sites/ianpalmer/2023/09/26/california-lawsuit-oil-companies-global-warming-climate-change-deception-what-the-jury-needs-to-know-part-1/?sh=3b12d01a1c4f.

Palmer, Ian. 2023f. "California Lawsuit: Oil Companies, Global Warming, Climate Change, Deception: What The Jury Has To Consider. Part 2." Forbes.com, September 29, 2023. https://www.forbes.com/sites/ianpalmer/2023/09/29/california-lawsuit-oil-companies-global-warming-climate-change-deception-what-the-jury-has-to-consider-part-2/?sh=32a3ec8866cc.

Palmer, Ian. 2023g. "Here's How Oil and Gas Companies Are Jumping on Hydrogen, but Is It Over-Rated." Forbes.com, June 8, 2023. https://www.forbes.com/sites/ianpalmer/2023/06/08/heres-how-oil-and-gas-companies-are-jumping-on-hydrogen-but-is-it-over-rated/?sh=7b5b07d932e3.

Palmer, Ian. 2023h. "Hydrogen Is Ramping up in the Energy Transition, but It May Be Oversold." Forbes.com, May 30, 2023. https://www.forbes.com/sites/ianpalmer/2023/05/30/hydrogen-is-ramping-up-in-the-energy-transition-but-it-may-be-oversold/?sh=1582b24a5f1a.

Palmer, Ian. 2023i. "Investment in Oil and Gas: Occidental Presentation Implies Strong Years Coming for Energy Security." Forbes.com, January 12, 2023. https://www.forbes.com/sites/ianpalmer/2023/01/12/investment-in-oil-and-gas-occidental-presentation-implies-strong-years-coming-for-energy-security/?sh=6c366a266abc.

Palmer, Ian. 2023j. "A Fresh Reading of Exxon's Predictions of Global Warming and Climate Change from 40 Years Ago." Forbes.com, January 17, 2023. https://www.forbes.com/sites/ianpalmer/2023/01/17/a-fresh-reading-of-exxons-predictions-of-global-warming-and-climate-change-from-40-years-ago/?sh=67a4db5a7840.

Palmer, Ian. 2023k. "The U.S. is Watching Battery Deployment and Success Down Under." Forbes.com, July 18, 2023. https://www.forbes.com/sites/ianpalmer/2023/07/18/the-us-is-watching-battery-deployment-and-success-down-under/?sh=2163ed725da0.

Palmer, Ian. 2023l. "Oil Companies and US Government are Beginning a Transformative Leap to Control Carbon Emissions." Forbes.com, September 12, 2023. https://www.forbes.com/sites/ianpalmer/2023/09/12/oil-companies-and-us-government-are-beginning-a-transformative-leap-to-control-carbon-emissions/?sh=4f378ed66ae7.

Palmer, Ian. 2023m. "No Room for Both in the Permian—Oil and Gas Versus Nuclear Waste." Forbes.com, May 21, 2023. https://www.forbes.com/sites/ianpalmer/2023/05/21/no-room-for-both-in-the-permian--oil-and-gas-versus-nuclear-waste/?sh=4ee5c26f4a1d.

Palmer, Ian. 2023n. "New Way to Get 700 C Degrees Heat and Electricity from Hydrogen—Without Burning It." Forbes.com, November 9, 2023. https://www.forbes.com/sites/ianpalmer/2023/11/09/new-way-to-get-700-c-degrees-heat-and-electricity-from-hydrogen--without-burning-it/?sh=d607c0b5b07a.

Palmer, Ian. n.d. "What Global Warming Can Do to Us–Part 3: Sea Level Rise Serious by the Year 2100." IanDexterPalmer.com. Accessed February 9, 2024. https://www.iandexterpalmer.com/sea-level/.

Parkinson, Giles. 2023a. "Liddell to Host Giant Battery After AGL and Akaysha Win Australia's Biggest Capacity Tender." Renew Economy, November 22, 2023. https://reneweconomy.com.au/liddell-to-host-giant-battery-after-agl-and-akaysha-win-australias-biggest-capacity-tender/.

Parkinson, Giles. 2023b. "Battery Storage Grabs Growing Share of Peak Supply in World's Most Renewable Grid." Renew Economy, December 13, 2023. https://reneweconomy.com.au/battery-storage-grabs-growing-share-of-peak-supply-in-worlds-most-renewable-grid/.

Parkinson, Giles. 2023c. "Quinbrook Eyes 16-Hour Batteries for Sun Cable in Massive Storage Deal With China's CATL." Renew Economy, November 8, 2023. https://reneweconomy.com.au/quinbrook-eyes-16-hour-batteries-for-sun-cable-in-massive-storage-deal-with-chinas-catl/.

Parkinson, Giles. 2023d. "AGL Adds Three More Big Batteries, Concedes Gas Demand Will Fall Even With Coal Exit." Renew Economy, June 16, 2023. https://reneweconomy.com.au/agl-adds-three-more-big-batteries-concedes-gas-demand-will-fall-even-with-coal-exit/.

Parkinson, Giles. 2023e. "Tesla Reveals Plan to be Australian Energy Retailer and 'Disrupt Traditional Incumbents.'" Renew Economy, November 16, 2023. https://reneweconomy.com.au/tesla-reveals-plan-to-be-australian-energy-retailer-and-disrupt-traditional-incumbents/.

Parkinson, Giles. 2023f. "Giant Melbourne Battery May Triple in Size with Plans for 12-Hour Flow Battery Storage." Renew Economy, December 1, 2023. https://reneweconomy.com.au/giant-melbourne-battery-may-triple-in-size-with-plans-for-12-hour-flow-battery-storage/.

Parkinson, Giles. 2023g. "CSIRO Says Wind and Solar Much Cheaper than Nuclear, Even with Added Integration Costs." Renew Economy, December 20, 2023. https://reneweconomy.com.au/csiro-says-wind-and-solar-much-cheaper-than-nuclear-even-with-added-integration-costs/.

Parkinson, Giles. 2024. "Not Enough Droughts and Hefty Network Charges: The Big Barriers to Grid Storage." Renew Economy, March 12, 2024. https://reneweconomy.com.au/not-enough-droughts-and-hefty-network-charges-the-big-barriers-to-grid-storage/.

Peters, Adele. 2022. "An Oil-Company Spinoff Wants to Help Build 70 Direct Air Capture Plants by 2035." Fast Company, June 7, 2022. https://www.fastcompany.com/90758711/an-oil-company-spinoff-wants-to-help-build-70-direct-air-capture-plants-by-2035.

Pielke, Roger, Jr. 2022. "Don't Believe the Hype." *The Honest Broker* (blog), December 19, 2022. https://rogerpielkejr.substack.com/p/dont-believe-the-hype.

Pielke, Roger, Jr. 2023a. "What the Media Won't Tell You about ... Wildfires." *The Honest Broker* (blog), June 8, 2023. https://rogerpielkejr.substack.com/p/what-the-media-wont-tell-you-about-783.

Pielke, Roger, Jr. 2023b. "U.S. Climate Extremes: 2023 Year in Review." *The Honest Broker* (blog), December 19, 2023. https://rogerpielkejr.substack.com/p/us-climate-extremes-2023-year-in.

Pielke, Roger, Jr. 2023c. "Unbelievable: U.S. Flood Damage." *The Honest Broker* (blog), November 7, 2023. https://rogerpielkejr.substack.com/p/unbelievable-us-flood-damage.

Pielke, Roger, Jr. 2023d. "The Alimonti Addendum." *The Honest Broker* (blog), August 26, 2023. https://rogerpielkejr.substack.com/p/the-alimonti-addendum.

Pielke, Roger, Jr. 2023e. "The Energy Transition Has Not Yet Started." *The Honest Broker* (blog), June 29, 2023. https://rogerpielkejr.substack.com/p/the-energy-transition-has-not-yet.

Pielke, Roger, Jr. 2024a. "Global Disaster Losses: 1990-2023." *The Honest Broker* (blog), January 12, 2024. https://rogerpielkejr.substack.com/p/global-disaster-losses1990-2023.

Pielke, Roger, Jr. 2024b. "Apples, Oranges, and Normalized Hurricane Damage." *The Honest Broker* (blog), February 26, 2024.

https://rogerpielkejr.substack.com/p/apples-oranges-and-normalized-hurricane.

Pitts, Pietro. 2023a. "Australia Retains Top LNG Exporter Spot, Says IGU Report." Hart Energy, July 18, 2023. https://www.hartenergy.com/exclusives/australia-retains-top-lng-exporter-spot-says-igu-report-205816.

Pitts, Pietro. 2023b. "US LNG: A Growing Market Presence." Hart Energy, December 12, 2023. https://www.hartenergy.com/exclusives/us-lng-growing-market-presence-207425.

Plumer, Brad. 2023. "In a U.S. First, a Commercial Plant Starts Pulling Carbon From the Air." *New York Times Online*, November 9, 2023. https://www.nytimes.com/2023/11/09/climate/direct-air-capture-carbon.html.

Pollack, Nicole. 2024. "The Big Battle Brewing over LNG Exports." Canary Media, January 23, 2024. https://www.canarymedia.com/articles/liquefied-natural-gas/the-big-battle-brewing-over-lng-exports.

Purtill, Corinne. 2022. "Nuclear Fusion Breakthrough in California Seen as Milestone Toward Clean Energy Future." *Los Angeles Times Online*, December 13, 2022. https://www.latimes.com/science/story/2022-12-13/nuclear-fusion-breakthrough-seen-as-milestone-toward-clean-energy-future.

Rapier, Graham. 2023. "Tesla Gigafactories: A Look at the Manufacturing Hubs and their Future." Business Insider, July 13, 2023. https://www.businessinsider.com/tesla-gigafactory.

Rathi, Akshat. 2024. "Davos Elite See Trump Among Hurdles for Green Growth in 2024." Bloomberg, January 19, 2024. https://www.bloomberg.com/news/articles/2024-01-19/davos-elite-see-trump-among-hurdles-for-green-growth-in-2024?embedded-checkout=true.

Ravishankara, A. R., Johan C. I. Kuylenstierna, Eleni Michalopoulou, Lena HöglundIsaksson, Yuqiang Zhang, Karl Seltzer, Muye Ru, Rithik Castelino, Greg Faluvegi, Vaishali Naik, Larry Horowitz, Jian He, Jean-Francois Lamarque, Kengo Sudo, William J. Collins, Chris Malley, Mathijs Harmsen, Krista Stark, Jared Junkin, Gray Li, Alex Glick, and Nathan Borgford-Parnell. 2021. *Global Methane Assessment: Benefits and Costs of Mitigating Methane Emissions.* United Nations Environment Programme.

https://www.unep.org/resources/report/global-methane-assessment-benefits-and-costs-mitigating-methane-emissions.

"Renewable Energy Hub in Australia." n.d. BP. Accessed February 7, 2024. https://www.bp.com/en_au/australia/home/who-we-are/reimagining-energy/decarbonizing-australias-energy-system/renewable-energy-hub-in-australia.html.

Renshaw, Jarrett, and Scott Disavino. 2022. "U.S. LNG Exports to Europe on Track to Surpass Biden Promise." Reuters, July 25, 2022. https://www.reuters.com/business/energy/us-lng-exports-europe-track-surpass-biden-promise-2022-07-26/.

Reyes, Ronny. 2024. "More NYC Fires Caused by Lithium-Ion Batteries from E-Bikes in 2 Months than in All of 2019: FDNY Chief." *New York Post Online*, March 3, 2024. https://nypost.com/2024/03/03/us-news/more-nyc-fires-caused-by-lithium-ion-batteries-in-two-months-than-in-all-of-2019-fdny-chief/.

Roston, Eric. 2024. "Five Climate Lessons From the $1.8 Trillion Race to Net Zero." Bloomberg, February 9, 2024. https://www.bloomberg.com/news/articles/2024-02-09/5-climate-lessons-from-the-1-8-trillion-race-to-net-zero.

Russell, Clyde. 2023. "Cheaper Spot LNG Prices Tempt Some Asian Buyers Amid Supply Gains." Reuters, April 12, 2023. https://www.reuters.com/markets/commodities/cheaper-spot-lng-prices-tempt-some-asian-buyers-amid-supply-gains-russell-2023-04-12/.

Rystad Energy. 2023. Energy Transition Marathon. May 2023.

Sahagún, Louis. 2023. "California Sues Five Major Oil Companies for 'Decades-Long Campaign of Deception' about Climate Change." *Los Angeles Times Online*, September 16, 2023. Updated September 17, 2023. https://www.latimes.com/california/story/2023-09-16/california-sues-five-major-oil-companies-for-lying-about-climate-change.

Schuler, Timothy A. 2023. "Who Will Manage the US Climate Retreat?" Bloomberg, December 12, 2023. https://www.bloomberg.com/news/features/2023-12-12/climate-migration-s-billion-dollar-question-who-manages-the-retreat.

Segrist, Sandy. 2024. "Biden Administration Hits the Brake on New LNG Export Projects." Hart Energy, January 26, 2024. https://www.hartenergy.com/exclusives/biden-administration-hits-brake-new-lng-export-projects-207932.

SEI (Stockholm Environment Institute). 2023. *Production Gap Report: Phasing Down or Phasing Up? Top Fossil Fuel Producers Plan Even More Extraction Despite Climate Promises.* UN Environment Programme. https://doi.org/10.51414/sei2023.050.

Sharma, Ashish, Conrad Wasko, Dennis P. Lettenmaier. 2018. "If Precipitation Extremes Are Increasing, Why Aren't Floods?" *Water Resources Research* 54, No. 11: 8,545–8,551. https://doi.org/10.1029/2018WR023749.

Sharma, Niharika. 2023. "Totalenergies Pauses Its Plan to Join Adani's $50 Billion Green Hydrogen Bet." Yahoo, February 9, 2023. https://finance.yahoo.com/news/totalenergies-pauses-plan-join-adanis-084000292.html.

Silverstein, Ken. 2023. "Chevron Drills Deep Into Carbon Capture, Hydrogen, and Solar Solutions." Environment + Energy Leader, March 14, 2023. https://www.environmentenergyleader.com/2023/03/chevron-drills-deep-into-carbon-capture-hydrogen-and-solar-solutions/#.

Smith, Christopher E. 2023. "Cost Looms over Hydrogen's Role in Energy Transition." *Oil & Gas Journal Online*, May 1, 2023. https://www.ogj.com/energy-transition/article/14293445/cost-looms-over-hydrogens-role-in-energy-transition.

Souter, David, Serge Planes, Jérémy Wicquart, Murray Logan, David Obura, and Francis Staub, eds. 2021. *Status of Coral Reefs of the World: 2020*. Global Coral Reef Monitoring Network, October 5, 2021. https://www.unep.org/resources/status-coral-reefs-world-2020.

"South Australia on Course for Net Zero Power by 2027." 2023. En:former, November 30, 2023. https://www.en-former.com/en/south-australia-on-course-for-net-zero-power-by-2027/.

"South Australia's Virtual Power Plant." n.d. Government of South Australia. Accessed February 8, 2024. https://www.energymining.sa.gov.au/consumers/solar-and-batteries/south-australias-virtual-power-plant.

Spector, Julian. 2024. "Texas Will Add More Grid Batteries than Any Other State in 2024. Canary Media, February 26, 2024. https://www.canarymedia.com/articles/energy-storage/texas-will-add-more-grid-batteries-than-any-other-state-in-2024.

SPE (Society of Petroleum Engineers). 2021. "SPE Live: What is the Role of Oil & Gas in Achieving Net-Zero?" YouTube video. Streamed live on August 23, 2021. https://www.youtube.com/watch?v=iSRRnushDEg.

Statista. 2022. "Leading Coal Exporting Countries Worldwide in 2021." Accessed March 20, 2024. https://www.statista.com/statistics/270952/global-hard-coal-exports-2009/#:~:text=In%202021%2C%20Indonesia%20exported%20some, million%20metric%20tons%20that%20year.

Statista. 2023a. "Annual Carbon Dioxide (CO_2) Emissions Worldwide from 1940 to 2023." Accessed March 19, 2024. https://www.statista.com/statistics/276629/global-co2-emissions/.

Statista. 2023b. "Leading Oil and Gas Companies Worldwide Based on Market Capitalization as of December 2023." Accessed March 20, 2024. https://www.statista.com/statistics/272709/top-10-oil-and-gas-companies-worldwide-based-on-market-value/.

St. John, Jeff. 2023. "How Many Jobs is the Inflation Reduction Act Spurring? A Lot." Canary Media, November 1, 2023. https://www.canarymedia.com/articles/clean-energy-jobs/how-many-jobs-is-the-inflation-reduction-act-spurring-a-lot.

St. John, Jeff. 2024a. "Long Wary of Batteries, New York's Now Poised to Go Big on Energy Storage." Canary Media, January 22, 2024. https://www.canarymedia.com/articles/energy-storage/long-wary-of-batteries-new-yorks-now-poised-to-go-big-on-energy-storage.

St. John, Jeff. 2024b. "This Report Puts a Price Tag on the Climate Impacts of US LNG Exports." Canary Media, February 27, 2024. https://www.canarymedia.com/articles/liquefied-natural-gas/this-report-puts-a-price-tag-on-the-climate-impacts-of-us-lng-exports.

Stapczynski, Stephen, Ruth Liao, and Anna Shiryaevskaya. 2024. "A $290 Billion Investment Cements Natural Gas's Relevance for Decades." Bloomberg, January 10, 2024. https://www.bloomberg.com/news/features/2024-01-11/natural-gas-boom-to-hit-warming-world-trying-to-quit-fossil-fuels.

Storrow, Benjamin. 2023. "U.S. Battery Installations Soared in 2022, Reshaping Power Grids." *Scientific American Online*, February 22, 2023. https://www.scientificamerican.com/article/u-s-battery-installations-soared-in-2022-reshaping-power-grids/.

"SunGas Renewables Selected to Supply Green Methanol for New Hydrogen and E-Methanol Facility in Australia." 2023. Sungasrenewables.com, November 10, 2023.

https://sungasrenewables.com/sungas-renewables-selected-to-supply-green-methanol-for-new-hydrogen-and-e-methanol-facility-in-australia/.

Supran, G., S. Rahmstorf, and N. Oreskes. 2023. "Assessing ExxonMobil's Global Warming Projections." *Science* 379, No. 6628. https://doi.org/10.1126/science.abk0063.

Szymczak, Pat Davis. 2023. "Ukraine War to Accelerate Energy Transition: BP 2023 Energy Outlook." *Journal of Petroleum Technology Online*, January 31, 2023. https://jpt.spe.org/ukraine-war-to-accelerate-energy-transition-bp-2023-energy-outlook.

Twidale, Susanna. 2023. "Global Fossil Fuel Production Plans Far Exceed Climate Targets, UN Says." Reuters. Accessed February 6, 2024. https://www.reuters.com/business/energy/global-fossil-fuel-production-plans-far-exceed-climate-targets-un-says-2023-11-08/.

Two Bit da Vinci. 2024. "Google's Billion Dollar Gamble: You Won't Believe What it is!" YouTube video, posted January 27, 2024. https://www.youtube.com/watch?v=y0UH0vLjchg.

U.S. Methane Emissions Reduction Action Plan. 2021. The White House Office of Domestic Climate Policy. https://www.whitehouse.gov/wp-content/uploads/2021/11/US-Methane-Emissions-Reduction-Action-Plan-1.pdf.

Valle, Sabrina, and Ron Bousso. 2022. "Oil Giants' Massive Profits Revive Calls for Windfall Taxes." Reuters. Accessed March 16, 2024. https://www.reuters.com/business/energy/wrapup-global-oil-giants-rake-massive-profits-third-quarter-2022-10-28/.

Valle, Sabrina. 2023. "Occidental CEO Sees Potential to License 1,000 Carbon Capture Plants." Reuters. Accessed February 6, 2024. https://www.reuters.com/business/energy/global-fossil-fuel-production-plans-far-exceed-climate-targets-un-says-2023-11-08/.

Vicente-Serrano, Sergio M., Dhais Peña-Angulo, Santiago Beguería, Fernando Domínguez-Castro, Miquel Tomás-Burguera, Iván Noguera, Luis Gimeno-Sotelo, and Ahmed El Kenawy. 2022. "Global Drought Trends and Future Projections." *Philosophical Transactions of the Royal Society A* 380, No. 2238. https://doi.org/10.1098/rsta.2021.0285.

Vorrath, Sophie. 2023. "South Australia Sees 'Doubling in Demand' for Its Low Cost Green Electricity." Renew Economy, May 31, 2023. https://reneweconomy.com.au/south-australia-sees-doubling-in-demand-for-its-low-cost-green-electricity/.

Vorrath, Sophie. 2024. "Big Batteries and Small Solar Star in 2023, but Big Wind Commitments Plunge." Renew Economy, March 13, 2024. https://reneweconomy.com.au/big-batteries-small-solar-star-in-2023-but-big-wind-commitments-plunge-to-zero/.

Wade, Will and Ferman, Mitchell. 2024. "US Shale Drillers Seek to Power Oil Patch With Small Nuclear Reactors." Bloomberg.com News, April 1, 2024. https://www.bloomberg.com/news/articles/2024-04-01/us-shale-drillers-seek-to-power-oil-patch-with-small-nukes?embedded-checkout=true

"Warren Buffett Adds to Occidental Petroleum Corp Stake." 2023. Yahoo Finance, October 26, 2023. https://finance.yahoo.com/news/warren-buffett-adds-occidental-petroleum-090436288.html.

Waterworth, David. 2023. "Assault of the Batteries." Clean Technica. Accessed February 8, 2024. https://cleantechnica.com/2023/06/23/assault-of-the-batteries/.

Wesoff, Eric. 2024. "The Biggest Solar-Plus-Storage Project in the US Just Came Online." Canary Media, January 25, 2024. https://www.canarymedia.com/articles/solar/the-biggest-solar-plus-storage-project-in-the-us-just-came-online.

White, Natasha, Stephen Stapczynski, and Eko Listiyorini, 2024. "Oil Giants Plan to Bury Massive Amounts of CO2 in Southeast Asia." Bloomberg Green, March 27, 2024. https://www.bloomberg.com/news/articles/2024-03-28/exxon-shell-chevron-stake-co2-storage-claims-in-southeast-asia.

"Why Exxon is Suing its Shareholders." 2024. ExxonKnews, February 6, 2024. https://climateintegrity.org/news/political-business-and-legal-voices-call-for-new-jersey-tohold-climate-polluters-accountable.July?category=news%7Cpress-releases.

Williams, Curtis. 2023a. "US October LNG Exports Climb to Second Highest Level on Record." Reuters, November 1, 2023. https://www.reuters.com/business/energy/us-october-lng-exports-climb-second-highest-level-record-2023-11-01/.

Williams, Curtis. 2023b. "US November LNG Exports Approach Record Levels Amid Higher Output." Reuters, December 1, 2023. https://www.reuters.com/business/energy/us-november-lng-exports-approach-record-levels-amid-higher-output-2023-12-01/.

Woods, Darren W. 2023. "Low-Carbon Hydrogen: Fueling Our Baytown Facilities and Our Net-Zero Ambition." ExxonMobil News, January 30,

2023. https://corporate.exxonmobil.com/news/viewpoints/low-carbon-hydrogen.

Worland, Justin. 2023. "Biden's Big Oil Truce." Magzter. Accessed February 6, 2024. https://www.magzter.com/stories/news/Time/BIDENS-BIG-OIL-TRUCE.

World Energy Council. 2022. "World Energy Trilemma Index 2022." Accessed February 6, 2024. https://trilemma.worldenergy.org/reports/main/2022/World%20Energy%20Trilemma%20Index%202022.pdf.

Wrightstone, Gregory. 2017. *Inconvenient Facts: The Science That Al Gore Doesn't Want You to Know*. Silver Crown Productions.

Zapantis, Alex, Noora Al Amer, Ian Havercroft, Ruth Ivory-Moore, Matt Steyn, Kiaoliang Yang, Ruth Gebremedhin, et al. 2022. *Global Status of CCS 2022*. Global CCS Institute. https://www.globalccsinstitute.com/resources/global-status-of-ccs-2022/.